Drug Dynamics for Analytical, Clinical, and Biological Chemists

DRUGS AND THE PHARMACEUTICAL SCIENCES

A Series of Textbooks and Monographs

Edited by

James Swarbrick

School of Pharmacy

University of North Carolina

Chapel Hill, North Carolina

Volume 13. ORPHAN DRUGS, *edited by Fred E. Karch*

Volume 14. NOVEL DRUG DELIVERY SYSTEMS: FUNDAMENTALS, DEVELOPMENTAL CONCEPTS, BIOMEDICAL ASSESSMENTS, *edited by Yie W. Chien*

Volume 15. PHARMACOKINETICS, Second Edition, Revised and Expanded, *Milo Gibaldi and Donald Perrier*

Volume 16. GOOD MANUFACTURING PRACTICES FOR PHARMACEUTICALS: A PLAN FOR TOTAL QUALITY CONTROL, Second Edition, Revised and Expanded, *Sidney H. Willig, Murray M. Tuckerman, and William S. Hitchings IV*

Volume 17. FORMULATION OF VETERINARY DOSAGE FORMS, *edited by Jack Blodinger*

Volume 18. DERMATOLOGICAL FORMULATIONS: PERCUTANEOUS ABSORPTION, *Brian W. Barry*

Volume 19. THE CLINICAL RESEARCH PROCESS IN THE PHARMACEUTICAL INDUSTRY, *edited by Gary M. Matoren*

Volume 20. MICROENCAPSULATION AND RELATED DRUG PROCESSES, *Patrick B. Deasy*

Volume 21. DRUGS AND NUTRIENTS: THE INTERACTIVE EFFECTS, *edited by Daphne A. Roe and T. Colin Campbell*

Volume 22. BIOTECHNOLOGY OF INDUSTRIAL ANTIBIOTICS, *Erick J. Vandamme*

Volume 23. PHARMACEUTICAL PROCESS VALIDATION, *edited by Bernard T. Loftus and Robert A. Nash*

Volume 24. ANTICANCER AND INTERFERON AGENTS: SYNTHESIS AND PROPERTIES, *edited by Raphael M. Ottenbrite and George B. Butler*

Volume 25. PHARMACEUTICAL STATISTICS: PRACTICAL AND CLINICAL APPLICATIONS, *Sanford Bolton*

Volume 26. DRUG DYNAMICS FOR ANALYTICAL, CLINICAL, AND BIOLOGICAL CHEMISTS, *Benjamin J. Gudzinowicz, Burrows T. Younkin, Jr., and Michael J. Gudzinowicz*

Other Volumes in Preparation

Drug Dynamics for Analytical, Clinical, and Biological Chemists

Benjamin J. Gudzinowicz
Burrows T. Younkin, Jr.
Development Group
Clinical Chemistry Section
Department of Pathology
Rhode Island Hospital
Providence, Rhode Island

Michael J. Gudzinowicz
University of Texas System Cancer Center
Science Park – Research Division
Smithville, Texas

MARCEL DEKKER, INC. New York and Basel

Library of Congress Cataloging in Publication Data

Gudzinowicz, Benjamin J.
Drug dynamics for analytical, clinical, and biological chemists.

(Drugs and the pharmaceutical sciences ; v. 26)
Bibliography: p.
Includes indexes.
1. Pharmacokinetics. 2. Drugs--Metabolism.
I. Younkin, Burrows, T., [date]. II. Gudzinowicz, Michael J. III. Title. IV. Series. [DNLM: 1. Drugs--metabolism. 2. Pharmacology, W1 DR893B v.24 / QV 38 G923d]
RM301.5.G83 1984 615'.7 84-17480
ISBN 0-8247-7239-3

MARCEL DEKKER, INC.
270 Madison Avenue, New York, New York 10016

Current printing (last digit):
10 9 8 7 6 5 4 3 2 1

PRINTED IN THE UNITED STATES OF AMERICA

Preface

The purpose of this monograph was to introduce to analysts performing therapeutic drug monitoring the many physiological and biochemical aspects related to drug dynamics or drug action. It became increasingly evident to us that many technicians and analysts working in the areas of clinical chemistry, biochemistry, analytical chemistry, and toxicology, as well as those in some pharmaceutical laboratories, have been generating volumes of drug test results with little or no appreciation for what is involved in a drug's disposition in the patient's body after its administration by the physician.

This monograph was written in a condensed version so as to elucidate the various aspects involved in this ever-growing area. It should be of value to the layman as well as the professional, and covers such subjects as drug administration (routes of entry), principles governing drug absorption and distribution, major pathways of and factors influencing biotransformation of drugs, the elimination of drugs, and the mechanisms of drug action. It is essentially a book on drug dynamics—a basic primer for those performing drug analyses.

Benjamin J. Gudzinowicz
Burrows T. Younkin, Jr.
Michael J. Gudzinowicz

Contents

Drug Dynamics for Analytical, Clinical, and Biological Chemists

Introduction

It is not the purpose of this book to delve deeply into various aspects of pharmacology such as biostatistics, drug administration, distribution, biotransformation (metabolism) and elimination, drug interactions, dose-response and dose-time relationships, and so on, since these subjects are more authoritatively presented in many fine reference books [1-31]. Nevertheless, these areas will be discussed briefly in order to place into proper perspective the problems associated with the analysis of drugs and their metabolites in biological fluids or media. A basic understanding of the underlying principles is essential for analytical, clinical, and biological chemists to assist them in the design and conduct of more efficient, meaningful research approaches to problem solving and also to make them more appreciative of the complexity of biochemical analyses in general. On the other hand, the tools by which some pharmacological problems can be solved have been discussed in a three-volume analytical series dealing specifically with gas chromatography (GC) [32], mass spectrometry (MS) [33], and the integrated GC-MS analytical system [34], these instrumental techniques serving as the basis for a seven-volume set of books by Gudzinowicz and Gudzinowicz [35-41] for the analysis of drugs and their metabolic products in varied biological media. As in pharmacology, their fundamentals must also be understood in order to utilize these analytical techniques to their fullest potential.

1

Drug Administration: Routes of Entry

Drugs may be administered into the body by either of two routes, parenteral or enteral. If administered parenterally, the gastrointestinal tract is bypassed, the commonest parenteral routes being intravascular, intramuscular, subcutaneous, topical (percutaneous), intradermal, inhalation, intra-arterial, and intravaginal. On the other hand, the administration of drugs by the enteral route is achieved via a sublingual, oral, or rectal process. Since the purpose of drug administration is to ensure an effective level of the therapeutic agent at its site of action, various factors, such as passage of drugs across body membranes, absorption, distribution of the therapeutic agent in the body, drug metabolism or biotransformation, and rate of excretion or elimination of drugs, influence the attainment of such a level within the human or animal system under study [6].

I. PARENTERAL ROUTES OF ADMINISTRATION

A. Intravascular Administration

Intravenous injection of drugs directly into the bloodstream has several advantages: the rapidity by which the drug can be circulated with a minimum of delay, and the ability to control the rate of drug administration to establish and maintain within rather well-defined narrow limits a plasma plateau via a continuous but controlled infusion. By reducing its rate or stopping the infusion, local, excessive drug concentrations can be drastically reduced. Permitting a more rapid drug action than can normally be achieved by either subcutaneous or intramuscular administration, the intravenous approach is at times more suitable for therapeutics that are less readily adsorbed from the gastrointestinal tract or tissue depots but more susceptible to possible destruction before appreciable absorption could occur (e.g., protein hormones, blood plasma, plasma substitutes, and whole blood) [1]. On the other hand, possible complications may result using the intravascular route: (a) anaphylactoid reactions resulting from the administration of a therapeutic agent to a sensitized person, (b) embolism by the inadvertent administration of particulate matter into the vein, (c) erythrocyte agglutination or hemolysis as a consequence of hypotonic or hypertonic solution injection, (d) infection by bacterial contaminants, and (e) the fact that once injected, the drug cannot be recalled.

B. Intra-Arterial Administration

In order to achieve a high local concentration in a specific area of the body before the onset of plasma volume dilution, drugs may be injected into the artery. The intra-arterial route is used principally for diagnostic therapeutics; it is rarely used otherwise except for specialized techniques in cancer chemotherapy.

C. Intramuscular Administration

More vascular and less sensitive than subcutaneous tissue and more tolerant to irritant suspensions or solutions, drug absorption is generally rapid when injected in the gluteal or deltoid regions of skeletal muscle. The rate of absorption is nevertheless dependent on many critical factors: (a) ionization of the drug, (b) lipid solubility of the drug, (c) vascularity of the site, and (d) solution osmolality, in addition to many other possible variables.

D. Subcutaneous Administration

The same factors that determine the therapeutic agent's rate of absorption from skeletal muscle influence its absorption from subcutaneous tissues. Its absorption can be retarded by incorporating a vasoconstrictor agent in the drug solution to be administered, by drug implantation in a solid pellet form under the skin, by vasoconstriction initiated by local cooling, by limb immobilization, or by tourniquet application adjacent to the injection site to impede or block superficially lymphatic flow and venous drainage.

E. Intradermal Administration

Small amounts of isotonic materials can be injected into the skin and, if properly administered superficially at a depth not exceeding several cell layers, a wheal is formed. This drug administration route is used for the introduction of antigens for skin tests and anesthetics to create an insensitive area for subsequent painless, deeper injections of a local anesthetic via a larger needle.

F. Topical (Percutaneous) Administration

Dermatological administration of drugs and their subsequent absorption through the skin barrier, which behaves qualitatively like a cellular membrane (the rate of penetration governed largely by their water-soluble ions, polar molecules, and lipid-water partition coefficients), requires that consideration be given not only to the pharmacological effect of the active agent but additionally to the physical form of the preparation (wet preparations, powders, shake lotions, emulsions, creams, or hydrophilic ointments, pastes, ointments, or direct drug application either as a simple solution, tincture, paint, suspension, or as crystals), which may itself have a therapeutic or adjuvant effect [20]. On the other hand, the pharmacological effects or actions resulting from the form of the topical medication may vary: (a) antipruritic, (b) astringent, (c) keratoplastic, (d) keratolytic, (e) antibacterial, (f) anti-inflammatory, (g) antifungal, or (h) parasiticidal. To facilitate drug penetration through the skin, research efforts have been directed toward the development of pharmacologically inert organic solvents to expand the usefulness of the percutaneous route for drugs with systemic actions [1]. Iontophoresis has also been used to drive by electrical gradients drugs into the skin after an ionized drug in solution is placed in an absorbent material on the skin in contact with an electrode. The drug ions are forced to migrate through the epidermis by applying a galvanic current to this electrode and another placed elsewhere on the body surface.

G. Drug Administration by Inhalation

Inhalation of drugs as aerosols (either liquid or solid particles) enter the circulation by diffusing across the permeable alveolar membranes. The degree of impaction (a term used to describe the aerosol particles in the respiratory tract) is rate dependent: that of sedimentation, diffusion, and inertial precipitation. For different parts of the respiratory tract, the extent of impaction can be determined for various particle sizes. Furthermore, the rate of drug absorption through the alveolar epithelium is strongly particle size dependent. Under certain circumstances, aerosol therapy (tranquilizers, diuretics, penicillin, glycosides, and digitalis) have been used effectively for systemic effects. However, the unrestricted use of aerosols containing isoproterenol and related bronchodilators can be fatal, as indicated by Speizer et al. [42], who noted a large increase in mortality from asthma in England during a 7-year monitoring period, the death rate paralleling the over-the-counter sales of these aerosols.

II. ENTERAL ROUTES OF ADMINISTRATION

A. Sublingual Administration

Like that of the rest of the alimentary canal, the epithelial lining of the mouth behaves as a lipidlike membrane barrier to the passage of drugs. Drugs in tablet form placed under the tongue (sublingual) between the buccal and gingival mucosa are absorbed from this site and enter the systemic rather than the portal circulation, its absorption being much more rapid than after the drug is swallowed.

B. Oral Administration

The most common drug administration route is via the mouth, it being painless, convenient, and economical. However, as noted by Schanker [43], the oral route of administration has certain liabilities or disadvantages: (a) gastric acid or digestive enzymatic destruction of some drugs; (b) precipitation or insolubility in gastrointestinal fluids of some drugs; (c) formation of drug-food nonabsorbable complexes; (d) variable absorption rates as a consequence of physiological factors such as gastric emptying time, gastrointestinal motility, and mixing; (e) poor effectiveness in an emergency situation if the rate of absorption is too slow; and (f) possible gastric mucosa irritation, with resultant nausea or vomiting. In contrast to a compound administered in aqueous solution, its absorption rate, if administered in solid form, is influenced by its rate of solution. Levy [44] has shown that its rate of solution depends on several factors: (a) particle size, solubility, salt form, and crystallinity of the drug; (b) the disintegration rate of the solid drug in the gastrointestinal lumen; and (c) gastrointestinal pH, motility, and food content.

C. Rectal Administration

In the form of suppositories or, less commonly, by enema, drugs may be administered rectally either for systemic effects or as local therapy. As in other parts of the alimentary canal, drugs are transported across the rectal mucosa via the same mechanism. Possessing a rather small surface area (being devoid of villi but having a rich lymphatic and vascular supply), drug absorption is usually slow. For this reason, rectal medication is normally used when the oral route is unsatisfactory, especially when orally administered drugs are either destroyed in the intestinal fluids or cause gastric irritation. The inert substances used for suppository preparations include polyethylene glycol, cocoa butter, and glycerinated gelatin.

2

Principles Governing Drug Absorption and Distribution

The principles governing drug absorption across biological membranes are the same as from the gastrointestinal lumen: the degree of drug ionization, lipid solubility, and molecular size. The extent to which a molecule will be transported from one biological compartment to another separated by a membrane structure is greatly influenced by these physicochemical properties, but as pointed out by Rowland [45], there are other critical determinants of its absorption when the drug is in solution that must be equally considered: (a) the chemical composition of the membrane it must cross, (b) the vascularity at and blood flow to the site, (c) the surface area exposed to the solution, and (d) the contact time of the drug at the absorptive site. When dissolution becomes the rate-limiting step in drug absorption, the solubility of the drug, its salt form, the surface area available for dissolution, the nature and volume of the vehicle, and other pharmaceutical factors become additional determinants. In Fig. 1 are illustrated the factors that influence drug absorption and disposition as well as the onset, duration, and intensity of drug effects. One notes that the drug must dissolve before being absorbed and that it passes across many lipoid barriers and some metabolizing systems before reaching the site of action.

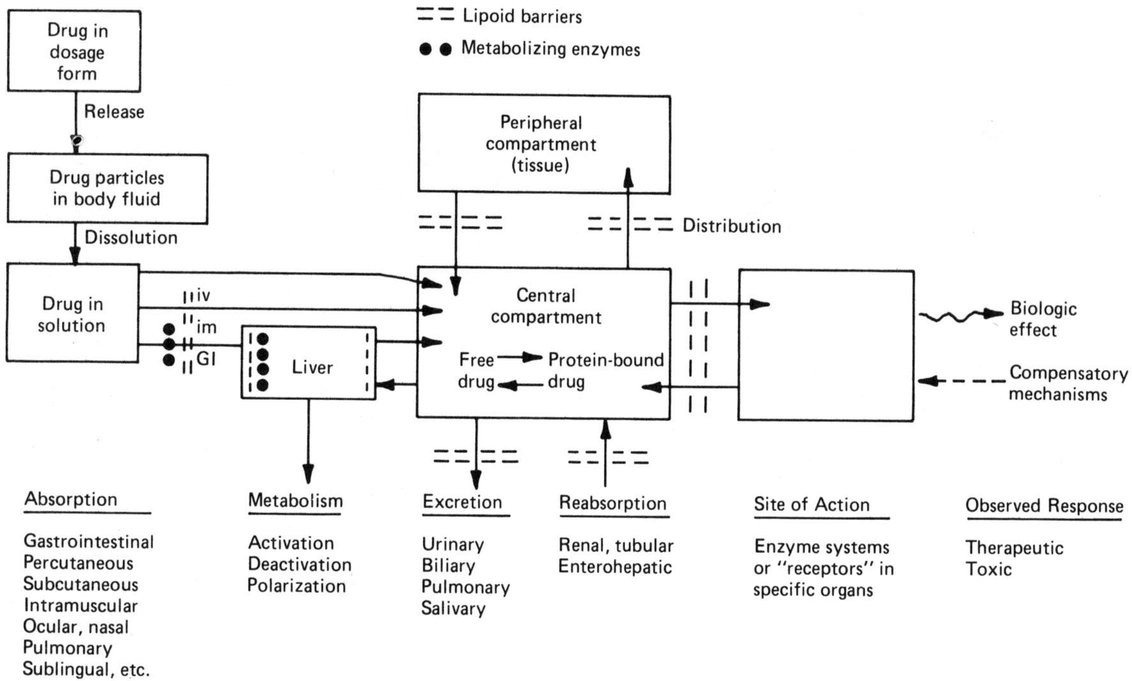

Figure 1 Diagram illustrating the factors that influence the onset, duration, and intensity of drug effects. Adapted from Barr [46], courtesy of the American Journal of Pharmaceutical Education.

Table 1 Common Acidic and Basic Drugs and Drug Classes

Weak acids	Weak bases
Anticoagulants (warfarin)	Amphetamines
Barbiturates	Antihistamines
Clofibrate (Atromid-S)	Erythromycin
Diphenylhydantoin (Dilantin)	Guanethidine (Ismelin)
Penicillins and cephalosporins	Major tranquilizers (phenothiazines)
Phenylbutazone (Butazolidin) and oxyphenbutazone (Tandearil)	Opiates (codeine, etc.)
	Quinidine
Sulfonamides and derivatives (thiazide diuretics and sulfonylurea hypoglycemic agents)	Tricyclic antidepressants (imipramine)
Salicylates	

Source: Adapted from Robinson [47], with the permission of Postgraduate Medicine.

I. PHYSICOCHEMICAL PROPERTIES

A. Ionization

Since most drugs are either weak acids or bases, their ionization is promoted in solution depending on the pH of the fluid or tissue. Whereas the dissociation of weak acids is facilitated by an alkaline pH environment, the reverse is true for weak-base drugs. Furthermore, the extent of their ionization markedly influences their lipid solubility; that is, the higher lipid solubility possessed by the non-ionized drug molecule passes more readily through most membrane barriers than its ionized species. Thus weak-acid drugs are often readily absorbed directly from the stomach, since they are nonpolar in the acidic gastric environment, whereas weak-base drugs are not significantly absorbed until they reach the alkaline environment of the small bowel [47]. Robinson [47] has categorized some common drugs and drug classes as to their acidic or basic natures as shown in Table 1. The influence of pH on polarity (ionization) is readily observed in Fig. 2.

Drug type	Acid (low pH) medium	Alkaline (high pH) medium
Weak acid	R-COOH	$R\text{-}COO^{\ominus} + H$
Weak base	$R\text{-}NH_3^{\oplus}$	$R\text{-}NH_2$

Figure 2 Influence of pH on polarity (ionization) or weak-acid and weak-base drugs. Adapted from Robinson [47], courtesy of Postgraduate Medicine.

Since the degree of ionization at a given pH as reflected in the dissociation constant pK_a (the negative logarithm of the acidic dissociation constant) determines the bond strength between the proton and base, the ionization of weak acidic drugs can be depicted by

$$[HA] = [H^+] + [A^-] \tag{1}$$

where

$[HA]$ = concentration of the non-ionized species
$[H^+]$ = concentration of the hydrogen ions
$[A^-]$ = concentration of the ionized species

The relationship between the pK_a of the drug, the pH of its solution, and the concentration of ionized/non-ionized drug species is given by the Henderson-Hasselbalch equation:

$$pK_a = pH + \log \frac{[HA]}{[A^-]} \tag{2}$$

For weak bases,

$$pK_a = pH + \log \frac{[A^-]}{[HA]} \tag{3}$$

If Eq. (2) is rearranged to

$$\frac{[A^-]}{[HA]} = 10^{pH - pK_a} \tag{4}$$

a large change can be effected in the extent of ionization by a relatively small change in pH near the pK_a.

The pK_a for either a weak acid or a weak base may be determined analytically by (a) titration [48, 49] or (b) spectrophotometrically. In the latter case, this is made possible by the fact that absorption spectra of many compounds are pH dependent. For acidic drugs, this can be determined from the expression

$$pK_a = pH + \log \frac{E_A - E}{E - E_{HA}} \qquad (5)$$

where

E = extinction coefficient of a mixture of both at a given pH and at the same wavelength
E_A = extinction coefficient of the ionized species, A^-
E_{HA} = extinction coefficient of the non-ionized molecule, HA, at the same wavelength

B. Lipid Solubility

The rate at which a drug penetrates a membrane barrier is dependent to a large extent on the lipid solubility of the drug. Lipid-soluble non-ionized drugs with high partition coefficients are capable of penetrating a variety of body cells and tissues whose boundaries are lipoid in character. Therefore, for a molecule to be transported across a membrane, it must be "dissolved" in the membrane initially, the degree of dissolution being dependent on the lipid solubility of the drug molecule. The presence of lipophilic (hydrophobic) nonpolar groups in the drug's structure markedly influences lipid solubility. Increased alkyl group chain length, insertion of multiple groups into the drug's structure, replacement of oxygen by sulfur, or the addition of a halogen all tend to increase or enhance lipid solubility. In general, the greater the lipid solubility, the greater the drug's transmembrane movement rate. In its passage through a membrane barrier, a correlation exists between lipid solubility and the partition coefficient K of the drug. The more lipid soluble the molecule, the higher its partition coefficient. Hogben et al. [50] demonstrated a rough parallelism between various weak electrolytes' rates of absorption and the lipid-to-water partition ratios of their non-ionized form and Schedl and Clifton [51] established a similar relationship between lipid solubilities and absorption rates in their study of the intestinal absorption of steroid compounds. By determining the solvent-water partition coefficient for the non-ionized form of the drug using a strongly acidic and strongly alkaline aqueous phase for acidic and basic drugs, respectively, compounds may in general be ranked according to their partition coefficients in a particular solvent with an approximate correspondence to the rank order observed extended to different solvents. It has also been shown that the increase in biological activity passes through a maximum with increasing nonpolar alkyl group chain length, the increase in activity roughly paralleling the increase in lipid solubility and decrease in water solubility (partition coefficient).

In 1964, Hansch and Fujita [52] studied the effect of substituent groups on distribution between water and 1-octanol (nonpolar solvent). Having measured the distribution coefficients for the parent compound and fluoro derivative, a value π for the substituent $-CF_3$ group is determined, its magnitude being dependent on the logarithmic difference between the partition coefficients.

$$\pi_{CF_3} = \log P_{CF_3} - \log P_H \tag{6}$$

where

P_{CF_3} = partition coefficient of the trifluoromethyl derivative
P_H = partition coefficient of the unsubstituted parent compound

Taken in conjunction with the Hammett sigma (σ) values [53-56], the π values have been applied effectively in correlating chemical structure, physical properties, and biological activities [3].

Hammett [53] has shown that a substituent group in a given position in an aromatic ring will exert a constant effect on a compound's reactivity, depending on its inductive and mesomeric properties. Expressed in terms of a constant σ value (a measure of the electron donating or withdrawing power of the substituent), these constants are additive. Hence the reactivity of a doubly substituted molecule can be calculated simply from the arithmetic sum of the group σ values. Hammett σ values therefore offer a convenient way for determining the inductive and mesomeric effects of specific groups in aromatic systems and, as a consequence, many investigations have been made to establish possible relationships to infrared group frequency shifts or changes. Linear relationships between group frequency shifts and σ values have been reported for the NH_2 stretching frequencies of anilines, the out-of-plane CH deformation modes of aromatic rings, the OH stretching frequencies of phenols, and the carbonyl frequencies of acetophenones and similar compounds [57].

In relating the application of the π and σ substituent constants to biological activity, Hansch derived the equation

$$\log \frac{1}{C} = -k_1\pi^2 + k_2\pi + \rho\sigma + k_3 \quad (7)$$

where

C = concentration of drug required to produce the biological response
k_1, k_2, k_3 = constants for the system under investigation
ρ = reaction constant
π = substituent constant for solubility contribution
σ = substituent constant for electronic contributions

Lipid solubility can be determined experimentally by determining the partition of the drug between an organic, immiscible, lipophilic solvent and an aqueous phase. After equilibrium is attained following vigorous mixing of the selected solvent with a buffered aqueous solution containing a weighed quantity of drug, the solvent/aqueous concentration ratio is calculated from the drug concentration found in each phase. Since the partition of the non-ionized species is pH independent, Butler [58] established its value from

$$\log\left(\frac{C_a}{C_o} - 1\right) = pH - pK_a \quad (8)$$

where

C_a = partition coefficient of the non-ionized species
C_o = observed or determined partition ratio

By determining the partition coefficient at two different pH levels and solving two simultaneous equations, both C_a and pK_a can be obtained. This route is suggested when the pK_a of the drug being studied is not known.

C. Molecular Shape and Size

In order for a drug to reach its site of action, it is subjected to a variety of biological processes prior to reaching its ultimate destination; all processes associated in some manner with the passage of drugs across body membranes. To achieve this objective, various transfer mechanisms (passive transfer or specialized transport) are involved in the movement of therapeutic agents across the body membrane. In the discussion of these transfer systems, it will be shown how molecular size and shape are an integral part of diffusion processes in general since diffusion coefficients are influenced by molecular size (or molecular volume) and molecular weight. For example, the diffusion coefficients of large molecules diffusing in water are inversely related to the cube roots of their molecular weights ($D \times M^{1/3}$ = constant, where D is the diffusion coefficient and M the molecular weight), whereas for small molecules, the diffusion coefficients are inversely related to the square roots of the molecular weights ($D \times M^{1/2}$ = constant). Furthermore, a relationship exists between permeability and molecular volume for small polar molecules. In the study by Ruhland and Hoffmann [59], the molecular volume was expressed in terms of molecular refraction (MR), which has volume dimensions

$$MR = \frac{M(n^2 - 1)}{d(n^2 + 2)} = M_V\left(\frac{n^2 - 1}{n^2 + 2}\right) \quad (9)$$

where

d = density of the compound
n = refractive index of the compound for a specified wavelength and temperature
M_V = molar volume of the compound = M/d

In 1933, Collander and Barlund postulated a lipoid-sieve plasma membrane concept for the permeation of plasma membranes by nonionized organic molecules. They correlated the cell permeability with the partition coefficients of compounds as a measure of lipid solubility. In their study, they noted an increase in permeability with decreasing molar volume in addition to a decrease in permeability with increasing lipid solubility [60].

II. DRUG TRANSFER SYSTEMS

For a drug to reach its receptor, it is not unusual for it to pass through several or more membrane barriers. To achieve this, its passage may be accomplished by one of several of the following transport mechanisms [6]:

1. Passive transfer
   a. *Simple diffusion*: characterized by the movement of drug molecules across the membrane whose rate of transfer is proportional to the concentration gradient on both sides of the membrane; its transport is accomplished without cellular expenditure of energy. Although only slightly

sensitive to temperature variations, the transfer is neither inhibited by other substances nor saturable.

b. *Filtration*: a process that essentially permits the greater portion of the bulk solvent and substances dissolved in it (but excluding constituents of large size) to flow through a porous membrane (e.g., the glomerular membrane of the kidney).

2. Specialized transport

a. *Active transport*: characterized by the movement of drugs and biological materials against a concentration or electrochemical gradient.

b. *Facilitated diffusion*: a special form of carrier transport system that operates along a concentration gradient of the permeating solute. No apparent energy is expended; it is not inhibited by metabolic poisons that interfere with energy production; at equilibrium, the solute will have the same concentration on either side of the membrane as in simple diffusion.

c. *Pinocytosis*: refers to the active transport of specific materials across a membrane by vacuolization (the ability of cells to engulf small droplets).

The movement of drugs across membranes can be expressed mathematically by Fick's diffusion equation. Since a concentration gradient does exist, the rate of transfer or diffusion dQ/dt depends on the concentration gradient $(C_2 - C_1)$ across the membrane, the area A of the membrane exposed to the passing solute molecules, and the partition coefficient K of the diffusing species between the membrane and the aqueous phase on either side of the membrane. The diffusion coefficient D is defined as the number of moles of drug that diffuse across a membrane of unit area in unit time when the concentration gradient is unity.

$$\frac{dQ}{dt} = \frac{KDA\ (C_2 - C_1)}{d} \tag{10}$$

where d is the thickness of the membrane.

Since there is a relationship between the diffusion coefficient and the radius of the molecule or particle,

$$D = \frac{RT}{N_0 6\pi r\eta} \tag{11}$$

where

R = gas constant
T = absolute temperature

N_0 = Avogadro's number
π = 3.1416
r = radius of the spherical particle
η = viscosity of the medium

it is possible to compare for purposes of approximation the diffusion coefficient and the molecular weight of the molecule. If one assumes that the molecule is a sphere with radius r and density d, the weight of a single molecule will be equal to $4/3\pi r^3 d$ while that of N_0 molecules will be equivalent to the molecular weight M.

$$M = \frac{N_0 4}{3\pi r^3 d} \tag{12}$$

Using Eqs. (11) and (12) and eliminating r, one obtains

$$D = \frac{RT}{N_0 6\pi\eta}\left(\frac{3M}{N_0 4\pi d}\right)^{-1/3} \tag{13}$$

It follows that D should vary inversely as the cube root of M, or

$$\frac{\Delta \log D}{\Delta \log M} = -\frac{1}{3} \tag{14}$$

In terms of permeability, Fick's law may be written

$$\frac{dQ}{dt} = PA(C_2 - C_1) \tag{15}$$

where P (permeability coefficient) = DK/d, depending on the diffusivity of the compound through the membrane, the partition coefficient, and the thickness of the membrane at the site of diffusion.

If a pH gradient exists across the membrane, the degree of drug ionization will vary on each side of the membrane, resulting in an unequal total concentration of the drug; thus a concentration gradient is created. Cohn [61] has shown that a simple general equation to describe the concentration ratios of a drug on both sides of the membrane at equilibrium can be derived as follows:

By rearrangement of Eq. (4), the concentration of the ionized species may be expressed in terms of [HA].

$$[A\;\,] = [HA]10^{pH-pK_a} \tag{16}$$

If C_1 and C_2 are the total concentration of drug (the sum of the ionized and non-ionized molecular species) on opposite sides of the membrane, these can then be mathematically expressed as

$$C_1 = [HA_1] + [A_1^-] = [HA_1] + [HA_1]10^{pH_1-pK_a} \tag{17}$$

$$C_2 = [HA_2] + [A_2^-] = [HA_2] + [HA_2]10^{pH_2-pK_a} \tag{18}$$

At equilibrium, the ratio R of the concentration across the membrane will be

$$R = \frac{C_1}{C_2} = \frac{[HA_1] + [HA_1]10^{pH_1-pK_a}}{[HA_2] + [HA_2]10^{pH_2-pK_a}} \tag{19}$$

At equilibrium, $[HA_1] = [HA_2]$ (the membrane being permeable to the non-ionized species or molecules). Therefore, it follows that

$$R_{acid} = \frac{1 + 10^{pH_1-pK_a}}{1 + 10^{pH_2-pK_a}} \tag{20}$$

For a weak base,

$$R_{base} = \frac{1 + 10^{pK_a-pH_1}}{1 + 10^{pK_a-pH_2}} \tag{21}$$

The total concentration of drug (for both weak bases and acids) is greater in the compartment where its ionization is greatest.

III. KINETICS OF DRUG ABSORPTION

The rate of absorption is dependent on the manner in which the drug is administered (i.e., the steady state achieved by constant infusions or the kinetics of the rise and fall of drug levels in the plasma after a single dose) as well as solubility and other critical physicochemical properties of the drug. How the drug is absorbed is of practical importance since its absorptive rate in large measure determines the time delay or latent period between its administration and onset of action. The magnitude of the time delay can vary depending not only on drug solubility and mode of administration (solution or solid form) but by the circulation to the site of absorption, route of administration, and area of the absorbing surface.

A. Zero-Order Absorption Kinetics for Constant Infusions

The mathematics for the derivation of the equations to describe the steady state [1, 9], where the rate of input is constant (k_i) and the output rate is exponential (k_o), is a zero-order kinetic process, the amount or content of the drug (C_0) in the system being accumulated until the steady state is reached.

$$\xrightarrow{k_i} C_0 \xrightarrow{k_o} \tag{22}$$

Consequently, if the constant input rate is k_i, it follows that the output rate can be expressed by k_oC_0. At steady state, where the output and input rates are equal, C_0 will have increased to C_1 and, in this system, it can be expressed as

$$C_1 = \frac{k_i}{k_o} \tag{23}$$

It should be noted that any change in the output rate will alter the steady state, thus shifting it to another steady-state equilibrium. This shift will also occur if the output rate (or both) is altered. When this occurs, the time course of the shift to reach the new steady state or plateau level is determined solely by k_o. Whether k_i or k_o is altered, the rate constant for the shift is nevertheless equal to k_o. Hence the shift half-time is equal or identical to the half-time of the output process, $0.693/k_o$. As pointed out by Goldstein et al. [1], the plateau principle is fundamental if one is to understand constant infusions, drug accumulation, dosage regimens, chronic toxicity by drug accumulation, and enzyme induction and stabilization. They derived it in the following manner:

For a change in C_0,

$$\frac{dC_0}{dt} = k_i - k_oC_0 \tag{24}$$

which becomes at the initial steady state,

$$C_{1(1)} = \frac{k_{i(1)}}{k_{o(1)}} \tag{25}$$

If the "in" and "out" rate constants are changed to new values, then, at the new steady-state value $C_{1(2)}$,

$$\frac{dC_0}{dt} = k_{i(2)} - k_{o(2)}C_0 \tag{26}$$

and at the new steady state,

$$C_{1(2)} = \frac{k_{i(2)}}{k_{o(2)}} \tag{27}$$

To determine C_0 as a function of time after the shift, one differentiates Eq. (26) with respect to time so that

$$\frac{dC_0}{k_{i(2)} - k_{o(2)}C_0} = dt \tag{28}$$

which, when rearranged, becomes

$$\left(-\frac{1}{k_{o(2)}}\right)\left(\frac{d[k_{i(2)} - k_{o(2)}C_0]}{k_{i(2)} - k_{o(2)}C_0}\right) = dt \tag{29}$$

If integrated,

$$\ln[k_{i(2)} - k_{o(2)}C_0] = -k_{o(2)}t + c \tag{30}$$

The constant of integration, c, can be determined when $t = 0$ and $C_0 = C_1$; the equation for its determination is given by

$$c = \ln[k_{i(2)} - k_{o(2)}C_{1(1)}] \tag{31}$$

Finally, by substitution and taking antilogarithms, we have

$$e^{-k_{o(2)}t} = \frac{k_{i(2)} - k_{o(2)}C_0}{k_{i(2)} - k_{o(2)}C_{1(1)}} \tag{32}$$

Concerned with the shift from one steady state to another, the total shift may be expressed as $(C_{1(2)} - C_{1(1)})$ and the amount by which C_0 has varied from the initial steady state as $(C_0 - C_{1(1)})$. With this information, the fraction f of the total shift at any value of C_0 can be indicated by

$$f = \frac{C_0 - C_{1(1)}}{C_{1(2)} - C_{1(1)}} \tag{33}$$

or

$$C_0 = fC_{1(2)} + C_{1(1)} - fC_{1(1)} \tag{34}$$

If this value of C_0 is substituted into Eq. (32) and $k_{o(2)}C_{1(2)}$ is substituted for its equivalent value, $ki_{(2)}$, then

$$\frac{C_{1(2)} - fC_{1(2)} - C_{1(1)} + fC_{1(1)}}{C_{1(2)} - C_{1(1)}} = e^{-k_{o(2)}t} \tag{35}$$

If simplified, it can be reduced to

$$1 - f = e^{-k_{o(2)}t} \tag{36}$$

As pointed out by Goldstein et al. [1], the shift kinetics are the exponential kinetics of a process with a rate constant $k_{o(2)}$. The half-time of the output process, $t_{1/2}$, is readily found when f is set to 0.5.

$$\ln 0.5 = -k_{o(2)}t \tag{37}$$

and

$$t_{1/2} = \frac{-\ln 0.5}{k_{o(2)}} = \frac{0.693}{k_{o(2)}} \tag{38}$$

B. First-Order Absorption Kinetics After Single-Dose Administration

To observe the initial rise and then the decay of drug levels after a single-dose administration, first-order kinetics can be used to describe drug absorption. Rowland [45] noted that initially, the total amount of drug is located at the absorption site, at which time the rate of absorption is at its maximum. Thereafter, the exponential decline or decay in absorption is accompanied by an increase in drug body levels and drug elimination rate. When the rate of absorption is equivalent to that for elimination, a maximum drug body level will have been attained. Thereafter, a decline in body drug levels will be noted. The faster the absorption, the faster the attainment of peak drug levels. On the other hand, the concentration of drug in blood or tissues may never reach therapeutic levels if the absorption is too slow. Furthermore, absorption differences may result as a consequence of (a) the route of administration or (b) the type of dosage form used even though the same dose is administered.

If D_0 is the dose of the drug introduced at the administration site when t = 0 and k_a is the absorption rate constant, it is possible to determine the drug remaining at the point of drug administration by the expression

$$D_r = D_0 e^{-k_a t} \quad (39)$$

where D_r is the amount of drug remaining after time t.

If D_{ac} is the amount of drug absorbed into the circulation and k_e is the elimination rate constant,

$$\frac{dD_{ac}}{dt} = k_a D_r - k_e D_{ac} \quad (40)$$

and if we substitute for D_r, we obtain

$$\frac{dD_{ac}}{dt} + k_e D_{ac} = k_a D_0 e^{-k_a t} \quad (41)$$

Teorell [62] has shown that for the case $k_a \neq k_e$, the following solution is obtained for the first-order differential above:

$$D_{ac} = \frac{k_a D_0}{k_e - k_a} \left(e^{-k_a t} - e^{-k_e t}\right) \quad (42)$$

However, when $k_a = k_e$, Eq. (42) reduces to

$$D_{ac} = k_e D_0 t e^{-k_e t} \quad (43)$$

The derivations of these equations are given in detail by Goldstein et al. [1] and for the purpose of graphing, Eq. (42) is rearranged to

$$D_{ac} = \frac{1}{(k_e/k_a) - 1} \left(e^{-k_e t/(k_e k_a)} - e^{-k_e t}\right) \quad (44)$$

so that D_{ac}/D_0 (vertical scale) may be plotted against $k_e t$ (horizontal scale).

By differentiation, it is possible to obtain the time at which the drug level is maximum. For $k_a \neq k_e$,

$$t_{max} = \frac{2.3}{k_a - k_e} \log \frac{k_a}{k_e} \tag{45}$$

For $k_a = k_e$,

$$t_{max} = \frac{1}{k_e} \tag{46}$$

Equations (45) and (46) indicate that, regardless of the dose administered, the time in which the maximum drug level is achieved is constant (the same).

Goldstein et al. [1] showed that the maximum amount of drug in the body could be estimated by substituting values of t_{max} into the original equations. For example, when $k_a \neq k_e$,

$$\frac{D_{ac}^{max}}{D_0} = \left(\frac{k_a}{k_e}\right)^{k_e/(k_e - k_a)} \tag{47}$$

When $k_a = k_e$,

$$\frac{D_{ac}^{max}}{D_0} = 0.368 \tag{48}$$

which says that the peak body level is only 37% of the dose.

IV. DISTRIBUTION OF DRUGS IN THE BODY

Once the drug is injected or absorbed into the bloodstream, it can either enter or pass through various body fluid compartments—plasma, interstitial fluid, transcellular fluids, the cellular fluids [63]—as well as tissues, as shown in Fig. 3. The distribution is determined by the transport of the therapeutic agent across various complex biological membranes. Of the water that constitutes nearly 70% of the human body, more than half is found within the cell membrane (intracellular water) and of the remaining extracellular water (Fig. 3), nearly 7% of the body weight is categorized as inaccessible water in the bone, whereas about 3% is transcellular water (water in the cavities of the intestine and stomach, in the content of the lumen of the kidneys and excretory glands, in the tracheobronchial tree, in the anterior chamber of the eyes, and in the cerebrospinal fluid), 7% as the water of blood plasma and the remainder as interstitial water [12]. The principal

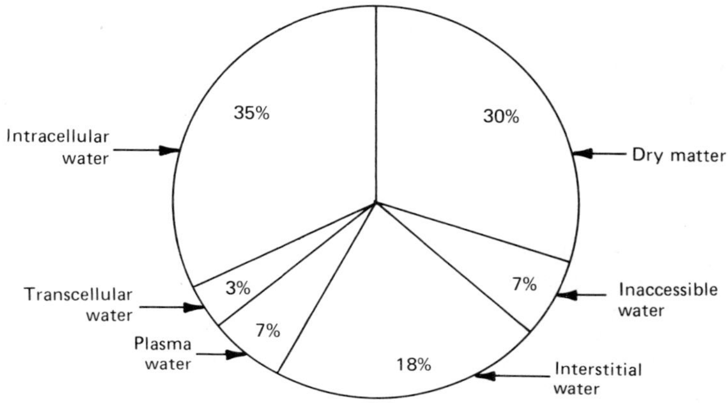

Figure 3 Approximate distribution of the various compartments in the body (compartments expressed as percentages of the total body weight). Adapted from Csaky [12], courtesy of Appleton-Century-Crofts.

compartments (plasma, interstitial, and intracellular) are separated by the capillary wall and cell membranes, respectively, is depicted in Fig. 4. With this compartmental arrangement, it is readily understood that a drug which fails to pass through the capillary wall is therefore distributed within the plasma, whereas those drugs that pass through the capillaries but cannot penetrate cell membranes remain in the extracellular compartment. On the other hand, the entire body water contains those drugs which permeate all membranes. From Fig. 4, we note that plasma proteins and protein-bound drugs remain within the plasma water since their passage through capillaries is restricted. This can be used diagnostically because if a drug is solely distributed in the plasma, it is a sign of complete binding to plasma proteins, while those which do not bind to plasma proteins can be used for the determination of the extracellular space. Lipid-soluble drugs are dispersed in the total body weight.

A. Apparent Volume of Distribution

The apparent volume of distribution V_d is defined as the volume of fluid in which the total amount of drug appears to distribute itself with a concentration equal to that observed in the blood. It assumes that the body acts as a single homogeneous compartment with respect to the blood.

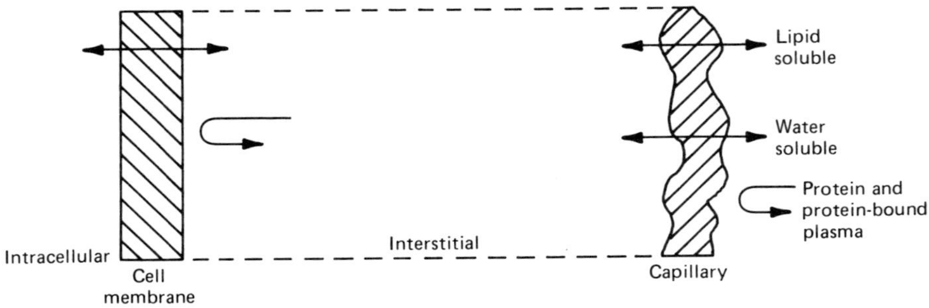

Figure 4 Distribution of the drug depends on binding to plasma proteins and on lipid solubility. Adapted from Csaky [12], courtesy of Appleton-Century-Crofts.

$$V_d = \frac{Q}{C_p} \tag{49}$$

where

V_d = apparent volume of distribution, liters
Q = total amount of drug in the body, mg
C_p = concentration of drug in the plasma, mg/liter

As pointed out by Butler [64]:

The apparent volume of distribution is a mathematical ratio that may or may not have a literal anatomical meaning in that it corresponds to an actual compartment of the body to which the drug is confined. . . . Implicit in the concept of volume distribution is the assumption that its value remains approximately constant as the values of Q and C_p vary over wide ranges. This is more likely to be true if the distribution of a drug corresponds to one of the water compartments than if a high proportion of the drug is localized outside plasma water. . . . Furthermore, the apparent volumes of distribution of acidic and basic drugs are influenced by metabolic and respiratory acidosis and alkalosis and may change quickly to significant degrees in response to changes in extracellular and intracellular pH. . . . However, the apparent volume of distribution as derived from plasma concentrations alone is not of itself sufficient proof that a drug is actually uniformly distributed in an anatomical compartment equal to that volume of distribution.

Substantiation by analysis of tissues would be required for conclusive establishment of the distribution.

In contrast to pharmacokinetic studies for drug characterization, which are normally limited to nonvolatile compounds, Chiou [65] derived theoretical equations to calculate the apparent volume of distribution and pulmonary clearance from inhalation of volatile or gaseous compounds based on plasma or blood data obtained after peripheral intravenous infusion. The analytical data thus obtained are valuable to an understanding of absorption, disposition, and potential toxicity of a large number of commonly used volatile or gaseous compounds.

Chiou's proposed equations were based on Perrier and Gibaldi's definition of the apparent volume of distribution V_d of a drug after having reached a pseudodistribution equilibrium state following intravenous administration [66], this being defined as

$$V_d = \frac{D_0}{kA_{ctc}} \qquad (50)$$

where

D_0 = dose of the drug administered
k = first-order rate constant obtained from the terminal exponential phase of the blood or plasma decay curve of a single or multiple exponential nature
A_{ctc} = area under the blood or plasma concentration versus time curve from time zero to infinity

As defined above, V_d can be used as a proportionality constant to relate either the blood or the plasma drug concentration to the total drug content in the body. However, Eq. (50) is valid only when there is insignificant drug elimination via exhalation of the intact therapeutic agent and/or biotransformation in the lung during its "first pass" in the pulmonary circulation before entering the left atrium, then the left ventricle, and finally the general circulation system. As noted by Rowland [67] in 1973, this first-pass effect is analogous to a drug's first-pass effect when administered orally, where a significant fraction of the total drug content is metabolized in the gastrointestinal wall and/or liver before reaching the general circulation.

Since most of the volatile or gaseous compounds are not metabolized significantly in the lung, Chiou therefore modified Eq. (50), assuming that the fraction of the intravenous dose not eliminated through exhalation in the first pass is f:

$$V_d = \frac{fD_0}{kA_{ctc}} \tag{51}$$

Whereas the total body clearance of the compound is equal to V_dk, the pulmonary clearance P_c is equivalent to

$$P_c = V_d k f_{ex} \tag{52}$$

or

$$P_c = \frac{f_{ex} fD_0}{A_{ctc}} \tag{53}$$

where

$$f_{ex} = \frac{(\text{total amount of compound exhaled}) - D_0(1 - f)}{fD_0} \tag{54}$$

As stressed by Chiou, f_{ex} is that fraction of the total compound entering the general circulation system which is exhaled via the lung, and for intravenously administered compounds that are exclusively eliminated by pulmonary exhalation, the value of f_{ex} should be 1.

In Eq. (51) the f value can be determined using, by analogy, the procedure adopted by Rowland [67] for the estimation of the extent of the first-pass effect after oral dosing; this is given by

$$f = 1 - \frac{P_c}{P_{bfr}} \tag{55}$$

where P_{bfr} is the pulmonary blood flow rate.

If Eq. (53) is substituted into Eq. (55), f may be expressed by the following relationship after an initial rearrangement is performed:

$$f = \frac{P_{bfr}}{P_{bfr} + (f_{ex}D_0/A_{ctc})} \tag{56}$$

If Eqs. (51) and (56) are combined,

$$V_d = \frac{P_{bfr}D_0}{k(A_{ctc}P_{bfr} + f_{ex}D_0)} \tag{57}$$

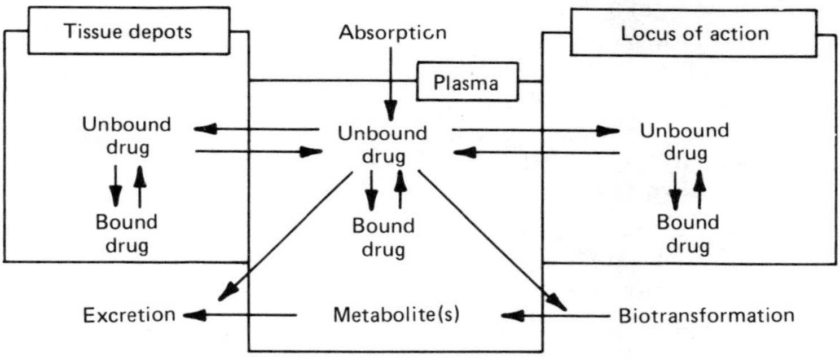

Figure 5 Diagram to illustrate how various factors influence the concentration of a drug in the plasma and hence its locus of action. Adapted from Gourley [70], courtesy of Butterworth.

As noted by Chiou, the pulmonary clearance [Eq. (55)] should be expressed in terms of blood volume cleared per unit of time so as to conform with the unit of pulmonary blood flow rate. The pulmonary blood flow can be estimated approximately, as it is equal to the cardiac output [68]. Spector [69] has compiled the cardiac output for various species.

B. Unequal Distribution of Drugs

Basically, the unequal distribution of drugs in the body is attributable to several factors: (a) binding to plasma protein, (b) cellular binding, (c) concentration in body fat, and (d) the blood-brain barrier [6]. Since relatively few molecules survive to reach the site of action, Fig. 5 depicts the predominant ways in which the drug may be "lost" rather than react at the normal site; collectively, these ways are categorized as "sites of loss." At the various sites of loss, drug depletion or loss may occur via a reversible storage depot process, an enzymatic system that produces metabolic alteration to a more or less active form of drug excretion before and after metabolism. Hence the competitive drug distribution between the sites of loss and action is dependent on the physicochemical properties of the drug: degree of ionization, solubility, and the nature and the strength of the forces binding the drugs at these sites [71].

1. *Drug Storage in Fat Depots*

Once administered into the body, the partition coefficient of a drug between neutral fat (a drug depot whose concentration relative to the total body weight may be as high as 50% in obese individuals and as low as 10% as a consequence of starvation) and extracellular water will determine its ultimate distribution. For example, highly lipid-soluble drugs (i.e., those with high lipid/water partition coefficients) will accumulate in these inert body fat depots where intermolecular attraction is chiefly that due to van der Waals forces. This will be reflected in a high apparent volume of distribution, thus confirming the possibility that a large proportion of the drug may be localized in the neutral fat.

Furthermore, there exists a relationship between the partitioning of lipid-soluble drugs between neutral fat and body water and the proportion of non-ionized drug present at physiological pH [70]. At pH 7.4, where hexobarbitone and other N-methylated barbituric acid analogs are completely non-ionized, these materials become highly concentrated or localized in fat partly because the substitution of a methyl group for the ionizable hydrogen attached to the nitrogen atom markedly decreases its acidity, which, in turn, increases lipid solubility due to the elimination of hydrogen bonding between the nitrogen and adjacent oxygen atoms.

Subtle structural changes will also have an effect on a drug's ability to partition rapidly into neutral fat. This has been dramatically demonstrated with thiopental (a thiobarbiturate with a pK_a of 7.6 which is nearly half-ionized in pH 7.4 plasma). The high lipid solubility for the undissociated molecule relative to its oxygen analog pentobarbital has been attributed to the presence of the sulfur atom. As a consequence, it has a short duration of anesthetic activity because of its decreased or depleted blood level concentration [72,73].

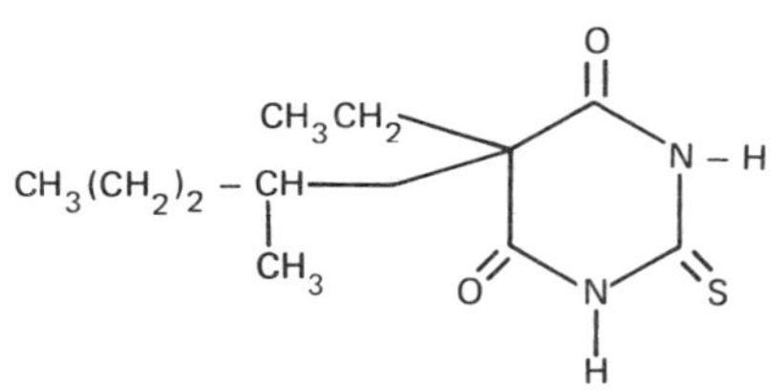

Thiopental

Axelrod et al. [74,75] have implicated lipid accumulation in the long duration of action of the adrenergic blocking agents dibenzyline and dibenamine. Of the initial dose, about 20% is localized in fat and, with time, it is slowly returned to the bloodstream. Agarwal and Harvey [76] have postulated that their activity of long duration may be attributed to the strength of the covalent bond formed at the active sites with these alkylating agents.

2. *The Blood-Brain Barrier*

Discussing some physicochemical factors in drug action, Brodie and Hogben [77] used the blood-brain barrier as a unique or singular example of unequal drug distribution because many intravenously injected drugs fail to enter the central nervous system, the cerebrospinal fluid, or the aqueous humor as readily as into other tissues. Behaving essentially as a barrier to nonlipid soluble substances, it has been shown by electron microscopic studies that the capillaries in the brain are enveloped by glial cells. This barrier permits fat-soluble substances that pass through these cellular membranes to permeate freely, in contrast to water-soluble materials, whose passage can be accomplished only if the glia has a specific transport mechanism for them.

Since drugs do cross epithelial cells into transcellular fluids, drug accumulation can occur; these fluids then function as transcellular depots of drugs. One such transcellular body depot is the gastrointestinal (GI) tract, where, as noted by Fingl and Woodbury [63], drug absorption in the GI fluid is solubility dependent. If it is relatively insoluble, slow absorption results and, under these circumstances, the GI tract may serve as a reservoir for maintaining the drug's plasma level as well as increasing its duration in the body. Because of the large differential between the two fluids, weak bases are passively concentrated in the stomach from the blood.

Since the cerebrospinal fluid contains no protein for drug binding, one can expect little drug accumulation in this fluid. The cerebrospinal fluid (CSF) differs from plasma water in several respects: that is, (a) the plasma is about 0.1 pH unit less acidic than CSF, (b) the CSF glucose level is lower due to the rapid metabolic utilization of glucose in CSF, (c) CSF has a higher Cl^- concentration, and (d) the K^+, Ca^{2+}, and PO_4^{3-} concentrations are greater in plasma than in CSF. The specific organ involved in maintaining the composition of the CSF is the choroid plexus, where the blood-cerebrospinal fluid barrier is located. Drugs may enter CSF via the choroid plexus or by diffusion directly across the capillaries into the interstitial fluid [1]. On the other hand, the routes of exit of drugs from CSF may differ from their entrance routes [63]. Regardless of lipid solubility or molecular size, all drugs and many endogeneous metabolites are transferred from the CSF into the venous blood sinuses through a system of large channels and valves in the arachnoid villi [78], the exit speed being the same for all substances and dependent on the bulk flow rate of cerebrospinal fluid. If the drug is lipid soluble, it can be eliminated by diffusion across the lipid portions of the CSF-blood boundary (i.e., by the same route it entered) or by diffusion into and out of the brain via the blood-brain barrier. Here its exit is aided by bulk CSF flow through the brain and into the capillaries [63].

Other transcellular fluids, such as aqueous humor (the fluid produced in the eye, which fills the anterior and posterior chambers in front of the lens and its attachments), endolymph, and joint fluids, are essentially drug-free, constituting only minor drug storage depots in the body. Finally, a major body storage depot for iodine is the luminal fluid of the thyroid, which is also capable of concentrating perchlorate as well as other monovalent anions [63].

3. *Plasma Protein Binding of Drugs*

Drug distribution in the body can be significantly altered by its ability to bind with proteins (in plasma, the protein usually responsible for binding is albumin, although globulins may participate in such protein-drug interactions, especially in relation to the binding of some hormonal agents and drugs) [6]. The drugs are reversibly bound to plasma albumin to varying degrees, as shown in Fig. 6. As listed by Davison [79] and Csaky [12], the importance or consequence of protein-drug binding lies in several areas:

1. Tissue distribution of the drug may be strongly modified.
2. In combination drug therapy, there is a competition between drugs for the binding sites, drugs with higher affinities displacing those with lower ones. If this situation occurs, the free (unbound) fraction and pharmacological effect of the displaced drug will be increased, as shown in Fig. 7. As illustrated by Robinson [47], the concentration of the active form may be doubled by as small an absolute change as 3% in binding equilibrium. Since acidic drugs tend to be anionic in the alkaline environment of plasma, they compete for binding sites on plasma protein. Figure 8 lists some clinically important acidic drugs that have been shown to interact. Of these, the sulfonylurea hypoglycemic agents and oral anticoagulants have the most serious potentiating interactions with other drugs. Levine [4] and Hansten [80] have written excellent books on drug actions and reactions and drug interactions, respectively, that of Hansten being concerned with the clinical significance of drug-drug interactions and drug effects on clinical laboratory results.

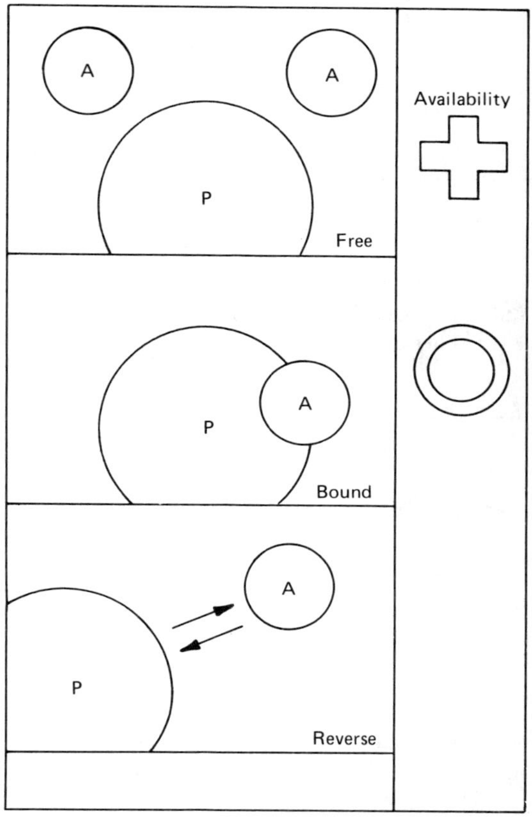

Figure 6 Drug distribution by transport in blood bound to plasma albumin. Bound fraction is inactive reservoir of drug. Adapted from Robinson [47], courtesy of Postgraduate Medicine.

3. Only the free fraction of the drug is in equilibrium with the extravascular compartments which are the site of the biophase. Hence this free concentration is available for distribution into the tissues only at the site of action.
4. Binding may be modified in disease states due to changes in nature and amount of protein, dehydration, or pH alteration.
5. Drug excretion may be delayed by prevention of tubular secretion or of glomerular filtration.

Drug	Bound	Free
A	97%	3%
A + B	94%	6%

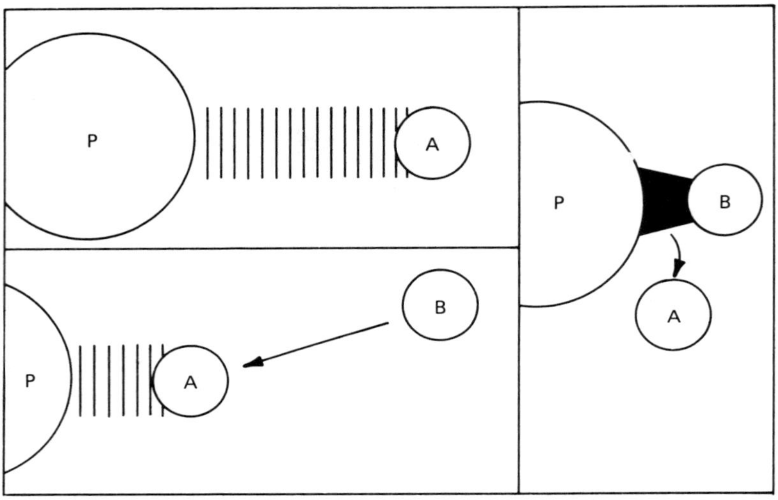

Figure 7 Competition for protein-binding sites. Drug B, with higher binding affinity, tends to displace drug A at shared binding sites, resulting in 3% absolute but 100% relative increase in concentration of free (active) form of drug A. Adapted from Robinson [47], courtesy of Postgraduate Medicine.

6. Since intestinal absorption of drugs is a diffusion process (its rate dependent on the concentration gradient on both sides of the epithelium), free drug absorption is enhanced since plasma protein binding of the drug increases the free drug's concentration gradient.
7. The relative "inherent potency" of drugs is often determined by plasma binding.
8. Accumulation is more likely to occur with a drug that is strongly bound to plasma proteins.

Since most binding reactions are both simple and reversible, Davison treated protein binding as a mass action problem such that

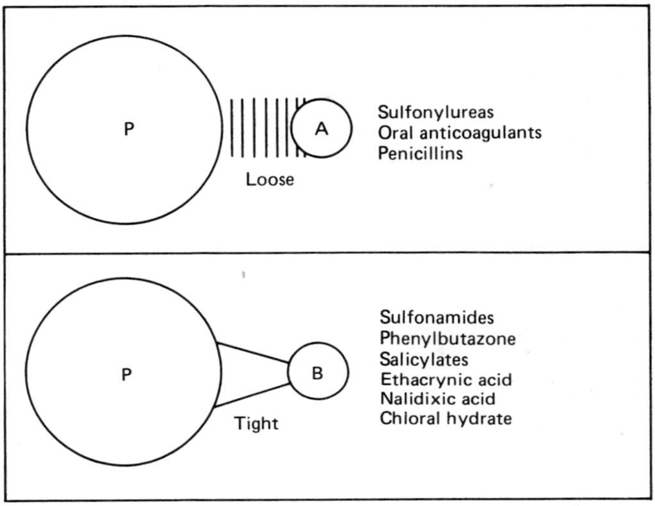

Figure 8 Some commonly used acidic drugs whose potentiation by another drug with high affinity may produce toxic effects. Adapted from Robinson [47], courtesy of Postgraduate Medicine.

$$(P_f) + (D_f) \rightleftharpoons (P_f D_f) \tag{58}$$

where

(P_f) = concentration of free protein
(D_f) = concentration of free drug
$(P_f D_f)$ = concentration of the protein-drug complex

This protein-drug complex will have at equilibrium an association constant K_A; this is defined as

$$K_A = \frac{(P_f D_f)}{(P_f)(D_f)} \tag{59}$$

From Eq. (59), the bound drug fraction β may be defined as

$$\beta = \frac{\text{moles of bound drug}}{\text{total moles of protein}} = \frac{(P_f D_f)}{(P_f D_f) + (P_f)}$$

$$= \frac{K_A (P_f)(D_f)}{K_A (P_f)(D_f) + (P_f)} = \frac{K_A (D_f)}{1 + K_A (D_f)} \tag{60}$$

It follows that for n independent but identical sites, a series of independent equations can be derived which, when summated, may be expressed as

$$\beta_1 + \beta_2 + \cdots + \beta_n = \beta_{total} = \frac{nK_A(D_f)}{1 + K_A(D_f)} \tag{61}$$

Often a given protein has more than one type of binding site (each with its own association constant), so that a more general equation is required to indicate this binding site variability. If such is the case, this situation can be expressed by

$$\beta_{total} = \frac{n_1 K_{1A}(D_f)}{1 + K_{1A}(D_f)} + \frac{n_2 K_{2A}(D_f)}{1 + K_{2A}(D_f)} + \cdots + \frac{n_n K_{nA}(D_f)}{1 + K_{nA}(D_f)} \tag{62}$$

In this form, Eq. (62) will yield a hyperbolic curve when plotted. Since the value of (D_f) can be measured and β_{total} calculated from the free and total drug concentration and protein concentration, Davison [79] showed that the data can be plotted in either of several ways in order to obtain values of n and K_A.

1. If β versus (D_f) is plotted directly, a hyperbolic plot results where $\beta = n$ at the plateau and at $\beta = n/2$, $(D_f) = 1/K_A$.
2. If plotted reciprocally, $1/\beta$ versus $1/(D_f)$, a straight line is given which can easily be fitted by statistical methods. If Eq. (61) is inverted and plotted,

$$\frac{1}{\beta} = \frac{1 + K_A(D_f)}{nK_A(D_f)} = \frac{1}{nK_A(D_f)} + \frac{1}{n} \tag{63}$$

   where one readily sees that the y intercept is given by $1/n$ and its slope by $1/nK_A$.
3. A straight line will result if $\beta/(D_f)$ versus β is plotted. As stressed by Davison, this approach is used most frequently for plotting binding data to obtain values for n and K_A.

By rewriting Eq. (61), we have

$$\beta + \beta K_A(D_f) = nK_A(D_f) \tag{64}$$

and rearranging Eq. (64) yields

$$\frac{\beta}{(D_f)} = nK_A - \beta K_A \tag{65}$$

where the y intercept is nK_A, the slope = K_A, and the x intercept is equal to n.

Davison notes that there are problems in this mathematical treatment of data, these arising from interactions between groups [81] and from electrostatic [82], osmotic, and competition effects. The methods used to determine protein-drug binding are varied: (a) equilibrium dialysis [83], (b) ultrafiltration [84], (c) gel filtration [85,86], (d) differential spectrophotometry [87,88], (e) changes in physical properties of drug solutions (i.e., estimating the degree of binding by measuring either the osmotic pressure, vapor pressure, electromotive force, conductivity, boiling point or freezing point), (f) bioassay, or (g) electrophoresis [89] and crossing electrophoresis [90].

It should be noted that the law of mass action and its subsequently derived equations to describe protein-drug binding can be also explained in terms of its dissociation constant K_D rather than K_A [1]. In this case, Eq. (59) would be given as

$$K_D = \frac{(P_f)(D_f)}{(P_f D_f)} \tag{66}$$

In this situation, if β is the number of moles of drug bound per total moles of protein, then

$$\beta = \frac{(P_f D_f)}{(P_f D_f) + (P_f)} = \frac{n(D_f)}{K_D + (D_f)} \tag{67}$$

If Eq. (67) is rearranged so that

$$\beta K_D + \beta(D_f) = n(D_f) \tag{68}$$

and then

$$\frac{\beta}{(D_f)} = -\frac{\beta}{K_D} + \frac{n}{K_D} \tag{69}$$

a straight line will result if $\beta/(Df)$ is plotted against β. The resulting graph is often called a "Scatchard plot" [91].

It is also possible to determine the free binding energy ΔF from the relationship [92]

$$\Delta F = -RT \ln nK_A = -RT \ln K_{A(tot)} \tag{70}$$

where

R = gas law constant
T = absolute temperature
n = number of binding sites on each protein molecule
K_A = association constant or the single binding constant
$K_{A(tot)}$ = total binding constant

Muller and Wollert [93] used the foregoing approach to observe the unusual high difference in the binding of d- and l-oxazepam hemisuccinate enantiomers to human and bovine serum albumin, presenting data regarding the affinities of the enantiomers to both albumins, the number of binding sites, and the degree of stereospecific binding.

On the other hand, Chiou and Hsiao [94] devised a new simple approach to study protein binding of volatile and gaseous compounds (fluorocarbon aerosol propellants, halothane, and cyclopropane) in an aqueous protein solution. It was based on the assumption that an increase in the partition coefficient of the compound between the protein solution and air phase, as compared to that between the aqueous phase not containing protein and air phase, is attributed to the binding of compound to the protein.

To determine the percentage of compound bound to the protein in an aqueous solution at equilibrium, they assumed that in a closed system the volatile or gaseous substance at equilibrium would be distributed between the headspace and protein solution such that its partition coefficient could be given by

$$P_c = \frac{C_p}{C_g} \tag{71}$$

where

P_c = partition coefficient of the compound
C_p = concentration of the compound in the protein solution
C_g = concentration of the compound in the headspace

Since the compound in the protein solution may exist in either the free or bound state, C_p may be substituted for C_t in Eq. (72).

$$\% \text{ Bound drug} = \frac{100C_b}{C_f + C_b} = \frac{100C_b}{C_t} \tag{72}$$

where

C_b = compound concentration in the bound form
C_f = compound concentration in the free form
C_t = total compound concentration ($C_f + C_b$) in the solution

If rearranged, Eq. (72) now becomes

$$\% \text{ Bound drug} = \frac{100\,(C_p - C_f)}{C_p} \tag{73}$$

If Eq. (73) is divided by C_g, it follows that

$$\% \text{ Bound drug} = \frac{100[(C_p/C_g) - (C_f/C_g)]}{C_p/C_g} \tag{74}$$

or

$$\% \text{ Bound drug} = \frac{100\,(P_c - P_f)}{P_c} \tag{75}$$

where P_f is the partition coefficient of the free drug form between the headspace and the protein solution.

From Eq. (71), P_c can be determined experimentally, whereas P_f must be estimated indirectly, by determining the free form concentration in the headspace at equilibrium with the same liquid system that does not contain the protein.

In their study, the partition coefficients were determined by gas chromatographic means; samples from both the headspace and liquid phase were analyzed by direct injection onto the GC column.

The partition coefficients of the propellants in the buffer or protein solutions were derived using the following relationship:

$$P_c = \frac{C_l}{C_g} = \frac{C_c V_c / f V_l}{C_g} = \frac{C_c V_c}{f C_g V_l} \tag{76}$$

where

C_l = propellant concentration at equilibrium in liquid solution

C_g = propellant concentration at equilibrium in the gas phase (headspace)
C_c = propellant concentration in cyclohexane extract
V_c = volume of cyclohexane used for extraction
V_l = volume of buffer or protein solution transferred for extraction
f = fraction of recovery

Since the propellant concentration is proportional to its gas chromatographic peak height, PH_c and PH_g can be substituted for C_c and C_g, respectively. If substituted in turn into Eq. (76), the following equation is obtained:

$$P_c = \frac{(PH_c)V_c}{fV_1(PH_g)} \qquad (77)$$

4. *Drug Binding by Other Body Constituents*

In addition to plasma protein, drug binding may occur with other body constituents or sites. For example, different chemical classes of drugs may interact with nucleic acids, and some evidence exists that drugs can be bound between adjacent base pairs of the DNA double helix [95-98]. Strong DNA interactions have been reported with drugs such as chloroquine and quinacrine which become highly localized within cells [99-101], whereas Potts [102] has shown that a number of polycyclic aromatics have a strong affinity in vitro with melanin in uveal pigment granules and with synthetic melanin obtained via dihydroxyphenylalanine oxidation. This was first noted using phenothiazines [103] and chloroquine [104].

Various amine-type compounds are electrostatically bound to acidic mucopolysaccharides, creating drug storage depots in connective tissue. In addition to the affinity of calcium for collagen, other cations chemically related to calcium are known to be deposited in both bone and other connective tissue, these cations being predominantly Mg, Sr, Ba, Be, and Ra.

Localization of certain drug compounds may also result from a variety of tissue protein-binding constituents as listed by Davison [79]: (a) cardiac glycosides and procaine bind to cardiac proteins, (b) thyroxine to mocrosomes of liver and muscle, (c) estradiol is taken up by the cytoplasmic and nuclear fraction of the endometrium, and so on.

Of equal and primary importance are the factors that must be considered that affect binding: (a) type of chemical compound [105], (b) type of protein, (c) patient's physiological and/or pathological condition, (d) variations existing between binding species [106, 107], as well as (e) human genetic differences [108].

3

Major Pathways of and Factors Influencing Biotransformation of Drugs

The metabolism or biotransformation of drugs is achieved by either of four types of specific processes or reactions: mechanisms based on oxidation, reduction, hydrolysis, or conjugation reactions. In turn, these chemical reactions have been classified as being either nonsynthetic (those involving oxidation, reduction, or hydrolysis which can result in activation, change in activity, or inactivation of the parent drug) or synthetic (those involving conjugation, the coupling between the parent drug or its metabolite, and an endogenous substrate that is usually a carbohydrate, an amino acid, derivatives of these, or an inorganic sulfate) [63].

As noted by Levine [4], the synthetic drug reactions or conjugations occur if the conjugating agent possesses an appropriate "center for conjugation," such as carboxyl (-COOH), amino ($-NH_2$), hydroxyl (-OH), or sulfhydryl (-SH). Although some therapeutic agents may not have these required reactive structural moieties (functional groups), one might be acquired by a nonsynthetic reaction; this nonsynthetic reaction end product or metabolite then undergoes further conjugative biotransformations. In humans, the number of synthetic pathways of biotransformation are limited, the predominant ones being methylation, glucuronide synthesis, acetylation, sulfate conjugation, and glycine and glutamine conjugations. Of special import is the fact that these synthetic chemical processes alter drugs and normal metabolites into

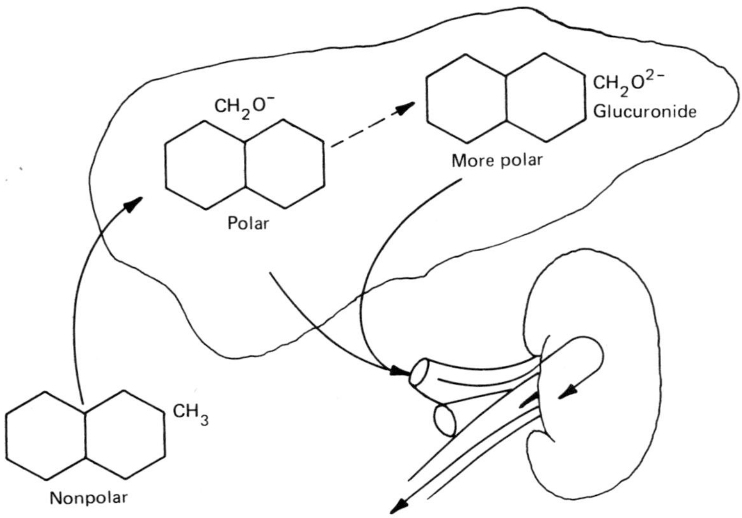

Figure 9 Hepatic drug metabolism. Drug molecules become more polar, favoring urinary excretion. Adapted from Robinson [47], courtesy of Postgraduate Medicine.

compounds that are, respectively, pharmacologically and biologically inactive. Being highly ionized products, the conjugated metabolites are less lipid soluble than the parent compound, thus decreasing their likelihood of being reabsorbed. Hence these "deactivated" drugs are more rapidly eliminated in the urine and feces. The more polar a compound becomes, the less lipid soluble it is, thus favoring urinary excretion, as depicted in Fig. 9. With given agents, one can predict the reactions that will occur but not their degree or extent of reaction.

On the other hand, Levine regards the nonsynthetic reactions (oxidation, reduction, hydrolysis, or a combination thereof) as only the first stage of biotransformation since the second stage embraces all the conjugation reactions of the metabolites formed in the nonsynthetic processes. The necessity for the "two-stage transformation" is to permit these nonsynthetic products to be eliminated from the body, this being possible only after conjugation. Unlike the conjugative reaction process, the nonsynthetic processes can convert an inactive drug into an active agent or change an active drug into another pharmacologically active compound. Levine notes that if the latter change takes place, the

Table 2 Comparison of Synthetic and Nonsynthetic Drug Biotransformations

Synthetic reactions	Nonsynthetic reactions
A. Type of reaction	
Determined by functional group	Determined by functional group
Limited number	Wide variety
Relatively predictable	Relatively unpredictable
B. Metabolite	
Almost always less lipid soluble	Usually less lipid soluble
Almost always pharmacologically inactive	May have less, equal, greater, or different activity
C. Reactions catalyzed by nonmicrosomal enzymes	
All except glucuronide conjugation	Most hydrolyses; some oxidations and reductions
No stimulation of rate of biotransformation by other drugs	No stimulation of rate of biotransformation by other drugs
D. Reactions catalyzed by microsomal enzymes	
Only glucuronide conjugation	Most oxidations and reductions; some hydrolyses
Rate of reaction stimulated by drugs	Rate of reactions stimulated by many agents

Source: Adapted from Levine [4], courtesy of Little, Brown and Co.

resulting metabolite may have (a) a qualitatively similar activity but be either less, equally, or more active; (b) a qualitatively different type of activity; or (c) a greater toxicity. The pharmacological activity manifested depends on the rate of change or transformation of an active drug to another active form. The similarities and differences between synthetic-nonsynthetic reactions are summarized in Table 2.

It must also be remembered that many drugs possessing several functional reactive groups which can be altered chemically

Table 3 Major Route of Biotransformation of Some Common Functional Groups

Hydroxyl (-OH)	
Alcohols (straight chain, i.e., aliphatic):	oxidation, glucuronide conjugation
Phenols (ring structure, i.e., aromatic):	glucuronide conjugation; sulfate conjugation; methylation
Carboxyl (-COOH)	
Aliphatic:	oxidation; glucuronide conjugation
Aromatic:	glycine conjugation; glucuronide conjugation
Amino ($-NH_2$)	
Aliphatic:	deamination (removal of amino group and formation of aldehyde); glucuronide conjugation
Aromatic:	acetylation; glucuronide conjugation; methylation
Aromatic rings:	hydroxylation

Source: Adapted from Levine [4], courtesy of Little, Brown and Co.

by enzymatic activity may be susceptible to several oxidative reactions, thus giving rise to a variety of end products. The most probable major reactions of some important functional groups are shown in Table 3. It is rare that a drug will form a double conjugate since one such adduct markedly increases the likelihood for its excretion, but several doubly conjugated drugs have been isolated [109].

Like other metabolic processes, the biotransformation of drugs is catalyzed by enzymes and for the most part the chemical changes that drugs undergo in the body occur principally in the liver, the intestines, and the kidneys. However, the majority of drugs are metabolized by the hepatic microsomal enzymes. Brodie et al. [110] have shown that metabolic transformations mediated by hepatic microsomal enzyme systems include deamination; N-dealkylation; O-dealkylation; oxidation of thioethers; hydroxylation of aromatic ring systems; aromatization of hydroaromatic compounds; oxidation of aldehydes and alcohols; reduction of the nitro and azo groups; conjugation with sulfuric, glucuronic, mercapturic, amino, and acetic acids; N- and O-methylations, hydrolysis of esters and amides; dehalogenation; and replacement of sulfur by oxygen [71].

On the other hand, inhibition of the enzymes participating in drug transformation can result in prolonging the duration and intensity of the pharmacological action of the drug, whereas stimulation of the enzyme may result in a lowering of the drug's pharmacological activity (duration as well as intensity).

Additionally, the major factors influencing drug metabolism include (a) genetic factors (pharmacogenetics); (b) age and sex differences; (c) species and strain differences; (d) nutritional status and diet; (e) hormones such as the adrenal hormones, thyroxine, and insulin; (f) disease, and (g) decreased pseudocholinesterase activity in persons with liver cell damage, in various cancer types, in persons injured with extensive second- or third-degree burns [111], and so on.

I. ENZYMATIC SYSTEMS FOR DRUG BIOTRANSFORMATION

The enzyme systems responsible for the majority of drug transformations are located in the hepatic endoplasmic reticulum, which is a network of lipoprotein tubules extending throughout the cytoplasm and continuous with the cellular and nuclear membranes. Under electron microscopic examination, some of the reticulum membranes appear to be rough surfaced, which is caused by small ribonucleoprotein particles (Palade granules) called ribosomes (specifically associated with protein synthesis), whereas the smooth-surfaced microsomes, consisting of lipoprotein alone, possess the enzymes that are capable of metabolizing drugs.

The microsomal enzymes are characterized by three main features, as noted by Csaky [12]:

1. Since substances with high lipid solubility can serve as substrates, it is highly probable that they are lipid receptors.
2. Although the microsomal enzymes lack specificity and are able to metabolize substances of widely diverse structure, the compound must be lipid soluble to participate in the catalysis reactions.
3. The liver microsomal drug-metabolizing enzyme (with few exceptions) are exclusively involved with the biotransformation of foreign compounds (i.e., drugs) and not with normal body metabolism.

The oxidative reactions are probably the most important nonsynthetic biotransformation reactions. These oxidative reactions in which enzymes are involved differ from the usual dehydrogenases of intermediary metabolism since they require reduced nicotinamide adenine dinucleotide phosphate (NADPH) and molecular oxygen for their action. Considered as a mixed function oxidase system, it has been postulated that NADPH reduces the component in the microsomes which reacts with molecular oxygen to form an "active oxygen" intermediate. This activated intermediate is then transferred to the drug by the overall reaction proposed by Gillette [112] in 1963:

(1) $NADPH + A + H^+ \rightarrow AH_2 + NADP^+$

(2) $AH_2 + O_2 \rightarrow$ "active oxygen complex"

(3) Active oxygen complex + drug $\rightarrow$ oxidized drug + A + H_2O

$$NADPH + O_2 + \text{drug} + H^+ \rightarrow NADP^+ + \text{oxidized drug} + H_2O$$

In this proposed sequence of reactions, NADPH is oxidized by a flavoprotein enzyme, NADPH-cytochrome c reductase, forming $NADP^+$ while transferring through the flavoprotein an electron to the oxidized form (A) of cytochrome P450 which has already interacted with the drug substrate. In the reaction above, AH_2 is the reduced form of cytochrome P450. The rate of drug biotransformation appears to be related to both the amount of drug-P450 complex and its rate of enzymatic reduction.

Liver microsomes also contain enzymes that are involved in the hydrolysis of amides and esters, the reduction of azo and nitro groups, and the formation of glucuronides from free amines, alcohols, acids, and phenols. Furthermore, there are various sites where enzymatic reactions may occur. For example, esterases are found in plasma which deesterify or hydrolyze certain drugs, whereas peptidases in erythrocytes, plasma, and many other tissues are involved in the biotransformation of the biologically active polypeptides [63]. The enzymes associated with sulfate conjugation can be found in the soluble fraction of liver cells, whereas the acetylating enzymes are present predominantly in the mitachondria of the liver and kidney. Additionally, the small intestine, thyroid gland, testes, and skin also participate to some extent in drug metabolism [70]. In contrast to the lipid-soluble substances oxidized by the specific hepatic microsomal enzymes, water-soluble foreign compounds may be oxidized by normal peroxidases, oxidases, and dehydrogenases of the body.

A. Enzymatic Inhibition

All enzyme inhibitions result from the interaction of some component of the enzyme system with the drug. In order for enzymes to exert maximal activity, various activating agents are required, such as apoenzymes (protein constituents), coenzymes (vitamin derivatives), cofactors (inorganic ions), and substrates. Malmstrom and Rosenberg [113] list three different mechanisms for initiating enzyme activation:

1. Formation of a catalytically active species formed by the interaction between a metallic ion and the enzyme:

$$E + M \xrightleftharpoons{K_{(E)(M)}} EM \tag{78}$$

$$EM + S \underset{k_2}{\overset{k_1}{\rightleftharpoons}} EMS \xrightarrow{k_3} EM + P \tag{79}$$

where

E = enzyme
M = metallic ion
S = substrate
P = product

Using the equations above, the reaction rate V can be expressed as

$$V = k_3\left(\frac{(E)}{1 + K_m/(S) + K_m/K_{(E)(M)}(M)(S)}\right) \tag{80}$$

where K_m is the Michaelis-Menten constant.

2. Substrate-metal complex formation, which, in turn, reacts with the enzyme:

$$S + M \xrightleftharpoons{K_{(S)(M)}} SM \tag{81}$$

$$E + SM \underset{k_2}{\overset{k_1}{\rightleftharpoons}} ESM \rightarrow E + PM \tag{82}$$

For Eq. (82), the rate of reaction is given by

$$V = k_3\left(\frac{(E)}{1 + K_m/(S) + K_m/K_{(S)(M)}(M)(S)}\right) \tag{83}$$

Equations (80) and (83) are identical, differing only in their respective dissociation constants.

3. Combination of the enzyme-substrate complex with the metal. Using this mechanism for enzyme activation, one notes that its rate of reaction is similar to Eq. (83).

On the other hand, enzyme inhibition occurs when any drug interferes with a substrate's access to active binding sites, the inhibition caused by drugs being either reversible or irreversible [114]. The inhibition is irreversible when it increases with time when the inhibitory drug is present in excess, while it is categorized as reversible when there exists an equilibrium between the inhibitory drug and the enzyme.

Webb [115], in his treatise on enzymes and metabolic inhibitors, shows that the enzymatic process can be initiated by a drug in essentially two ways: (a) by changing the affinity of the substrate S for the enzyme E or (b) by altering the dissociation rate constant of the enzyme-substrate complex.

Korolkovas [116] illustrates these concepts vividly using the following model for his mathematical derivations. Gillette [117] and Westley [118] have treated the mechanisms of enzyme inhibition in a similar manner.

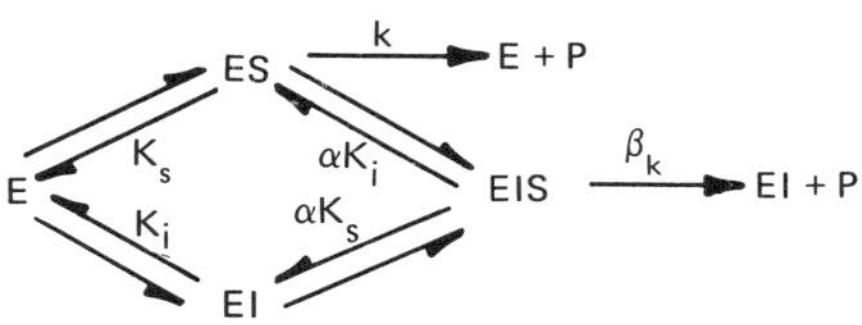

$$E + S \rightleftharpoons ES \xrightarrow{k} E + P \qquad K_S = (E)(S)/ES \tag{84}$$

$$E + I \rightleftharpoons EI \qquad K_i = (E)(I)/EI \tag{85}$$

$$EI + S \rightleftharpoons EIS \longrightarrow EI + P \qquad \alpha K_S = \frac{(EI)(S)}{(EIS)} \tag{86}$$

$$EI + I \rightleftharpoons EIS \longrightarrow EI + P \qquad \alpha K_i = \frac{(ES)(I)}{(EIS)} \tag{87}$$

where

K_i = constant of the inhibitory drug
K_S = dissociation constant for the enzyme-substrate complex (ES)
α = alteration in affinity
β = change in the dissociation rate of the complex induced by the inhibitory drug

I = inhibitory drug
P = dissociation product

Whether the drug acts as an activator or an inhibitor depends on the values of α and β.

By writing the relationships

$$(E_t) = (E) + (ES) + (EI) + (EIS) \tag{88}$$

$$v_i = k(ES) + k(EIS) \tag{89}$$

which can be combined with those given previously, Webb showed that

$$v_i = V_m\left(\frac{(S)[\alpha K_i + \beta(I)]}{(S)(I) + \alpha[K_i(S) + K_s(I) + K_sK_i]}\right) \tag{90}$$

where

v_i = enzymatic reaction rate in the presence of the inibitory drug
V_m = maximal rate, $k(E_t)$

The fractional inhibition f_i, which may be defined as $f_i = 1 - a$ (a being the fractional activity), can be expressed as

$$f_i = \frac{(I)[(S)(1 - \beta) + K_s(\alpha - \beta)]}{(I)[(S) + \alpha K_s] + K_i[\alpha(S) + \alpha K_s]} \tag{91}$$

Of the five types of inhibition that are distinguishable by considering the different values of α and β, only three main types are usually mentioned: (a) competitive, (b) noncompetitive, and (c) uncompetitive. Levine [4] defines a decrease in the metabolic rate of a given substrate as being competitive inhibition when the interfering agent is (a) a compound which is itself a substrate for the enzyme, or (b) a compound that undergoes no catalytic change but which combines reversibly with active sites of the enzyme by virtue of its structural similarity to the substrate. However, competitive inhibition can be overcome by a large enough concentration of substrate. On the other hand, noncompetitive inhibition describes a situation where metabolism is inhibited by an agent unrelated in structure to the substrate but one which is capable of combining with the enzyme so as to prevent enzyme-substrate complex formation. Uncompetitive or coupling inhibition is a particular form of mixed inhibition where the drug combines only with the enzyme-substrate complex at a site different from the one with which the substrate is combined.

Referring to Eqs. (90) and (91), a completely competitive inhibition results when the inhibitor completely prevents the combination of the enzyme with the substrate (i.e., when $\alpha = \infty$ and $\beta = 0$). This situation is described mathematically when

$$v_i = \frac{V_m(S)}{(S) + K_s[1 + (I)/K_i]} \tag{92}$$

$$f_i = \frac{(I)}{(I) + K_i[1 + (S)/K_s]} \tag{93}$$

When the inhibitor does not alter the binding of the substrate to the enzyme but prevents the dissociation of the ES complex, this describes a situation of complete noncompetitive inhibition ($\alpha = 1$ and $\beta = 0$).

$$V_i = \left(\frac{V_m K_i}{(I) + K_i}\right)\left(\frac{(S)}{(S) + K_s}\right) \tag{94}$$

$$f_i = \frac{(I)}{(I) + K_i} \tag{95}$$

Finally, when the inhibitor reduces the affinity of the enzyme for the substrate (thus preventing EIS complex dissociation), this is the case of mixed inhibition ($\infty > \alpha > 1$ and $\beta = 0$).

$$v_i = \left(\frac{\alpha K_i V_m}{(I) + \alpha K_i}\right)\left(\frac{(S)}{(S) + K_s\{[\alpha(I) + \alpha K_i]/[(I) + \alpha K_i]\}}\right) \tag{96}$$

$$f_i = \frac{(I)}{(I) + K_i\{[\alpha(S) + \alpha K_s]/[(S) + \alpha K_s]\}} \tag{97}$$

Uncompetitive inhibition, a form of mixed inhibition, occurs when α is less than 1.

As pointed out by Gillette [117], in most competitive inhibition studies, it is generally assumed that the inhibitor is not metabolized by the enzyme. However, using plotting techniques, it can be shown that the K_i value of the inhibitor is identical to its K_m

value if both the inhibitor and substrate are metabolized by the enzyme. In this case, if S_1 and S_2 are used to designate the substrate and inhibitor, respectively, their rate equations can be expressed as

$$v_1 = \frac{V_{m1}(S_1)/K_{m1}}{1 + [(S_1)/K_{m1}] + [(S_2)/K_{m2}]}$$

$$= \frac{V_{m1}}{1 + [K_{m1}/(S_1)][1 + (S_2)/K_{m2}]} \qquad (98)$$

$$v_2 = \frac{V_{m2}(S_2)/K_{m2}}{1 + [(S_1)/K_{m1}] + [(S_2)/K_{m2}]}$$

$$= \frac{V_{m2}}{1 + [K_{m2}/(S_2)][1 + (S_1)/K_{m1}]} \qquad (99)$$

where v_1, V_{m1}, and K_{m1} are used to describe the metabolism of the substrate whereas v_2, V_{m2}, and K_{m2} refer to the inhibitor's metabolism.

If the substrate and the inhibitor are converted to the same product, the total product rate of formation can be derived such that

$$V_t = v_1 + v_2 = \frac{[V_{m1}(S_1)/K_{m1}] + [V_{m2}(S_2)/K_{m2}]}{1 + [(S_1)/K_{m1}] + [(S_2)/K_{m2}]} \qquad (100)$$

Robinson [47] used Fig. 10 to illustrate an interfering drug which may either accelerate or inhibit the metabolism of another drug in the liver. He notes that many otherwise unrelated drugs that share the same metabolic reaction (e.g., glucuronidation) compete at this level and therefore may inhibit each other's metabolism. In humans, a number of drugs that act as inhibitors have been reported:

1. Bishydroxycoumarin and phenylbutazone inhibit the metabolic inactivation of tolbutamide.
2. Bishydroxycoumarin and disulfiram inhibit diphenylhydantoin metabolism, causing the anticonvulsant to accumulate to toxic levels.
3. SKF 525-A (diphenylpropylacetic acid diethylaminoethanol ester) inhibits the function of the microsomal oxidative

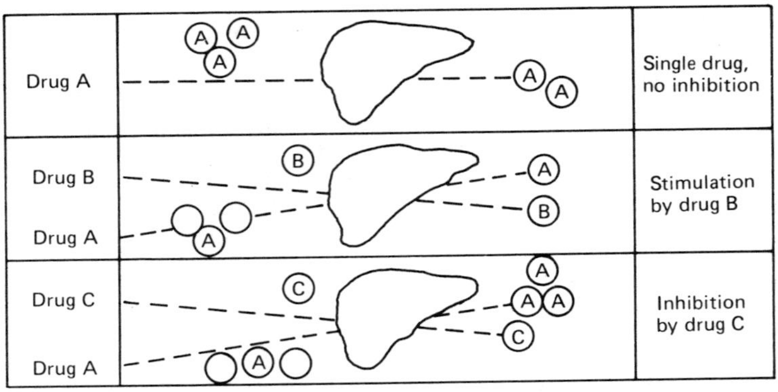

Figure 10 Drug interactions that alter liver metabolism. Drug B enhances metabolism of drug A by inducing hepatic microsomal enzymes and this reduces the effect of drug A. Drug C inhibits the metabolism of drug A by competition for metabolic pathways or by enzyme inhibition, resulting in excessive and toxic levels of drug A. Adapted from Robinson [47], courtesy of Postgraduate Medicine.

enzymes, in turn potentiating the action of those drugs which are primarily disposed of by the microsomal enzymes.

4. Morphine depresses the ability of the liver microsomes to N-dealkylate morphine and nalorphine.
5. Ethyl alcohol is inhibited in its metabolism by disulfiram (tetraethylthiuram disulfide).
6. Finally, such monoamine oxidase inhibitors as pargyline, iproniazid, and tranylcypromine block the metabolism of certain adrenergic amines in the mitochondria.

B. Enzymatic Stimulation

In contrast to enzyme inhibition by drugs, drugs can stimulate enzymatic activity, which, in turn, may accelerate the metabolism of another drug as shown in Fig. 10. In fact, the activity of the microsomal NADPH-dependent oxidative enzyme is enhanced by hundreds of compounds, a list that includes barbiturates, anticonvulsants, anesthetic gases, food additives, carcinogenics, CNS stimulants, steroid hormones, pesticides and herbicides, tranquilizers, hypoglycemic agents, antipyretics, muscle relaxants, and so on [4,12]. Sharing only one property common to all substances—lipid solubility—stimulation can result via two possible routes: (a) enzyme induction: increased rate of synthesis of new enzyme molecules, or (b) enzyme activation: an increase of the catalytic activity [12]. Table 4 lists some of the agents known to stimulate drug biotransformation in humans and some of the substances whose metabolism is affected [4].

Table 4 Agents Stimulating Microsomal Enzyme Activity in Humans and the Compounds Whose Biotransformation Is Affected

Stimulating agents (pretreatment)	Compounds whose rate of biotransformation is increased
Cigarette smoke	3,4-Benzpyrene (carcinogenic compound) Nicotine
Diphenylhydantoin	Cortisol
Glutethimide (Doriden)	Glutethimide (sedative-hypnotic) Warfarin
Griseofulvin	Warfarin
Meprobamate	Meprobamate (sedative)
Phenobarbital	Bilirubin (naturally occurring breakdown product) Bishydroxycoumarin (anticoagulant) Cortisol (naturally occurring hormone) Digitoxin (increases performance of the failing heart) Diphenylhydantoin (antiepileptic) Griseofulvin (antifungal agent) Phenobarbital (sedative) Testosterone (male sex hormone) Warfarin (anticoagulant)
Phenylbutazone (Butazolidin)	Aminopyrine (nonnarcotic analgesic) Cortisol Phenylbutazone (antirheumatic agent)
Tetrachlorodiphenylethane (DDD) (insecticide)	Cortisol

Source: Adapted from Levine [4], courtesy of Little, Brown and Co.

As a result of stimulated biotransformations, potential dangers and even unknown consequences can arise when a patient receives more than one drug. Hence the need of knowing the effect of each drug on the metabolism of others. Furthermore, since some drugs are self-stimulating (i.e., enhance their own metabolism), a drug may appear to be less toxic in the course of chronic toxicity studies. For this reason, drug concentrations in the plasma rather than the dose should be correlated with observed toxic effects. Finally, the administration of drugs in increased amounts or over prolonged periods of time may alter the normal metabolism of normal body substrates. In this regard, one must determine whether the acceleration of the metabolism of normal body substrates is beneficial or part of the toxic effects of drugs [60].

II. FACTORS AFFECTING DRUG BIOTRANSFORMATION

In addition to enzyme inhibitors and stimulators that influence drug metabolism, others are (a) genetic factors (pharmacogenetics), (b) species and strain differences, and (c) age and sex differences [6,12,70,71]. These aspects must be considered as pointed out by Daniels and Jorgensen [71], especially

> where there is more than one chemical pathway involved since the amount of metabolized product formed is determined by a rate process which is dependent on the concentration and activity of the enzymes responsible for the metabolism, and the accessibility to the enzymes and, thus, is subject to wide species and individual variation. The rate at which metabolism occurs has a direct relationship to the duration of action of a drug and varies with species and is also subject to individual variation within the same species.

A. Genetic Factors (Pharmacogenetics)

It is not too surprising that there is a variation of drug response in different subjects and that occasionally unexpected reactions are noted because, in humans, genetic heterogeneity between individuals has long been recognized. The mere absence of a gene required for the synthesis of a specific enzyme needed for normal metabolism results in an enzyme-deficiency situation, hence an inherent metabolic abnormality. In turn, abnormal drug reactions as a consequence of inherited diseases are rather well known. For example, Goth [6] cites the following: Barbiturates may produce severe attacks in congenital porphyria;

salicylates are dangerous in persons in whom glucuronyl transferase is congenitally absent (Crigler-Najjar syndrome); hypersensitivity to atropine occurs in patients suffering from Down's syndrome; and epinephrine or glucagon fail to produce hyperglycemia in persons who are deficient in the enzyme glucose-6-phosphatase (von Gierke's disease).

LaDu [119] classified pharmacogenetic abnormalities into several categories which are responsible for most genetically conditioned anomalous drug responses: (a) anatomic abnormalities, (b) drug metabolism disorders, (c) receptor site abnormalities, and (d) tissue metabolism disorders.

In addition to the glucose-6-phosphate dehydrogenase deficiency in red cells that is known to cause severe hemolytic episodes after the administration of oxidant drugs, isoniazid, an antitubercular drug, is inactivated primarily by acetylation, but also by other deactivation reactions, shown below:

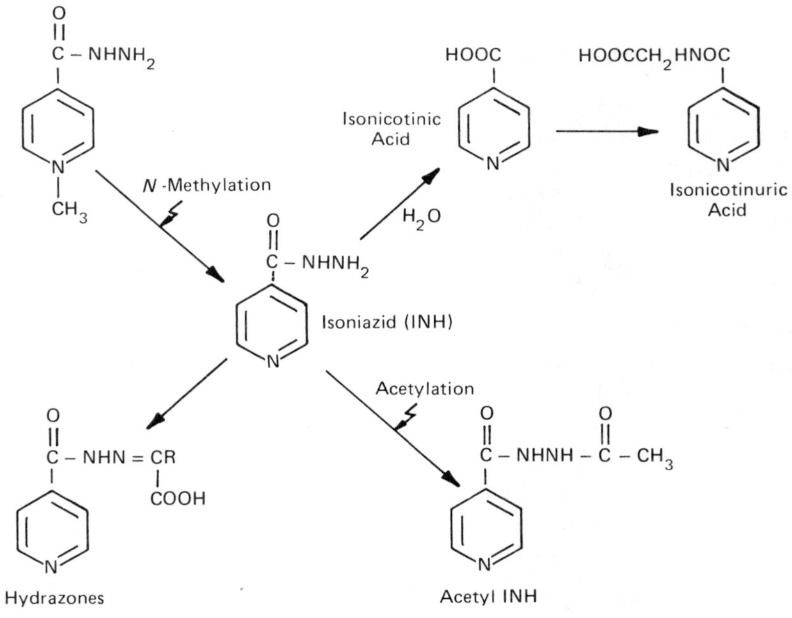

By analysis, it has been shown that following oral administration of the drug, the frequency distribution of the plasma concentration was bimodal rather than the usual bell-shaped or gaussian curve. Furthermore, when the percentage of "free" isoniazid appearing in the urine was determined in a large population sample, it was discovered that two classes could be distinguished with regard to its metabolism: slow inactivators and fast inactivators. Further studies revealed that both types are rather evenly distributed among American Caucasians and American Negroes, but the distribution of slow inactivators among Eskimos and Latin Americans was only 5 and 33%, respectively. When treated with isoniazid, patients who develop peripheral neuritis are usually slow inactivators and since the mechanism of toxicity is known, slow inactivators are normally given pyridoxine together with isoniazid. These variations in acetylation rate is attributed to the genetically controlled concentrations of acetylcoenzyme A.

Another class example is the enzymatic destruction of succinylcholine by plasma pseudocholinesterase. However, 1 patient in 3000 develops a prolonged muscular relaxation, which includes paralysis of the respiratory muscles. These persons were found to have the atypical form of pseudocholinesterase which has a very low affinity for succinylcholine.

B. Age Differences

In different species of experimental animals, there is a deficiency in the liver microsomal enzymes for biotransforming drugs into excretable polar metabolites [70,71]. Because newborn mammals are grossly deficient in enzymes that form glucuronic acid, which then undergoes catalytically induced glucuronide conjugation, it follows that premature infants and newborns cannot excrete drugs that are biotransformed to glucuronides or naturally occurring bilirubin. Hence drugs that are metabolized via this metabolic pathway will remain intact in the body for a relatively long time, where they may compete with bilirubin for binding to plasma albumin. Because of this binding competitiveness, sufficient bilirubin can be released, which can be extremely toxic.

C. Sex Differences

Although many experimental animals exhibit no sex variations in rate studies of drug metabolism, the rat was an exception, as noted by Gillette [120]. After puberty, the females showed significantly lower rates of metabolism than the males for a variety of drugs (barbiturates, sulfonamides, narcotics). After puberty, these differences have been attributed to androgen and related anabolic steroid production, which increases the enzymatic metabolizing activity of the liver microsomes, enzymatic activity paralleling rather closely the anabolic activity.

D. Species Differences

Although the metabolism of drugs may in many instances be common to most species (cats, dogs, mice, humans, etc.), certain biotransformation reactions may be totally lacking in others. For example, primary amines are not acetylated by dogs, whereas cats are not capable of forming phenolic glucuronides. Hens eliminate benzoic acid as ornithuric acid, but hippuric acid is excreted from dogs. Other major differences that have been recorded are [121]: (a) In rabbits, deamination of amphetamine occurs, but in dogs its aromatic ring is hydroxylated; (b) isomeric hydroxylated products of acetanilid result when administered to cats and dogs, hydroxylation occurring at the para position in dogs but at the ortho position in cats.

Csaky [12] used the following examples to illustrate differences in metabolic rates attributed to different species:

1. *Procaine*: Although rapidly hydrolyzed in humans by the plasma esterase, its hydrolytic rate is very slow in the horse.
2. *Barbiturates*: The metabolism of hexobarbital varies dramatically among species; its half-life in mouse, rat, dog, and humans being 19, 140, 260, and 360 min, respectively.
3. *Phenylbutazone*: Its half-life in varied species is: humans, 72 hr; monkey, 8 hr; dog, 6 hr; rat, 6 hr; horse, 6 hr; and rabbit, 3 hr.
4. *Meperidine*: In humans, its metabolic rate is nearly 17% per hour, whereas it is about 70% per hour in the dog.

E. Strain Differences

Rates of biotransformation of drugs not only vary among species as noted above, but variations exist from one strain to another within the same species. For this reason, Brodie [122] suggested that it would be extremely advantageous if a specific species and strain of experimental animal could be found which metabolizes a given class of drugs in the same manner as do humans.

F. Other Factors

Malnutrition and starvation depress drug metabolism. Furthermore, in experimental animals, drug biotransformations are also influenced by hormones such as the adrenal hormones, thyroxine, and insulin, and by disease. Liver cell damage lessens the activity of pseudocholinesterase. For instance, jaundice tends to depress the oxidative function and the glucuronic acid conjugation of liver microsomes.

III. PATHWAYS OF DRUG BIOTRANSFORMATION

Recall that drug-metabolizing reactions can be classified as nonsynthetic and synthetic. The synthetic reactions are those that conjugate drugs or their metabolites provided that specific molecular groupings are present in their chemical structure, whereas nonsynthetic reactions are those that convert the parent drug into active and inactive metabolites by oxidative, reductive, or hydrolytic means.

A. Nonsynthetic-Metabolizing Reactions

1. *Oxidative Reactions (Microsomal)*

Aromatic hydroxylation:

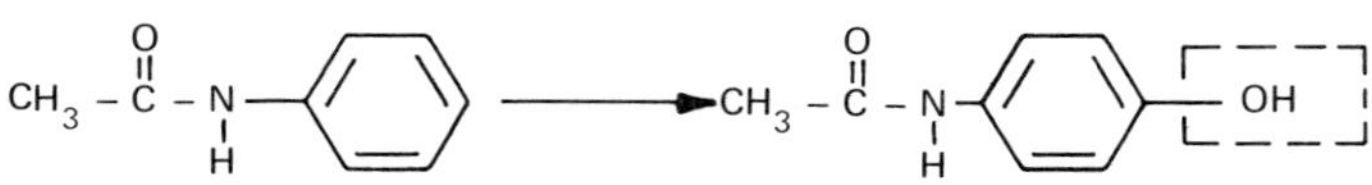

Acetanilide Acetaminophen

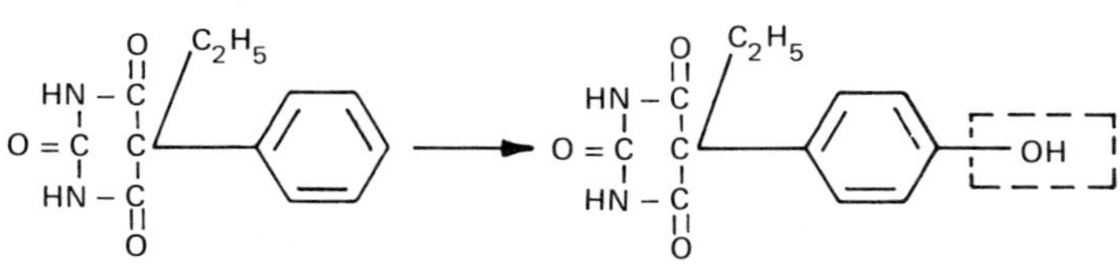

Phenobarbital *p*-Hydroxyphenobarbital

Aliphatic hydroxylation:

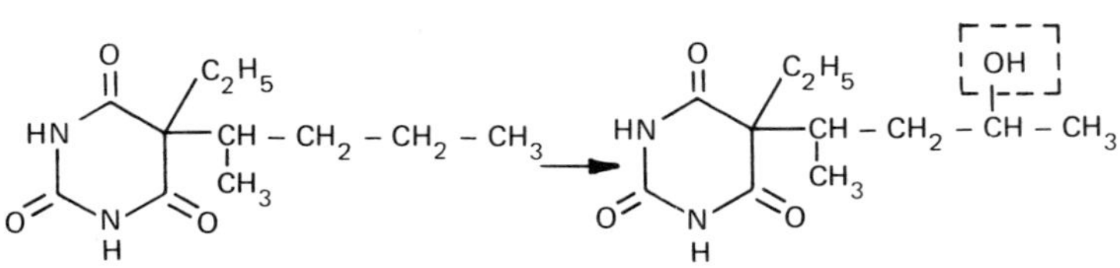

Pentobarbital Pentobarbital alcohol

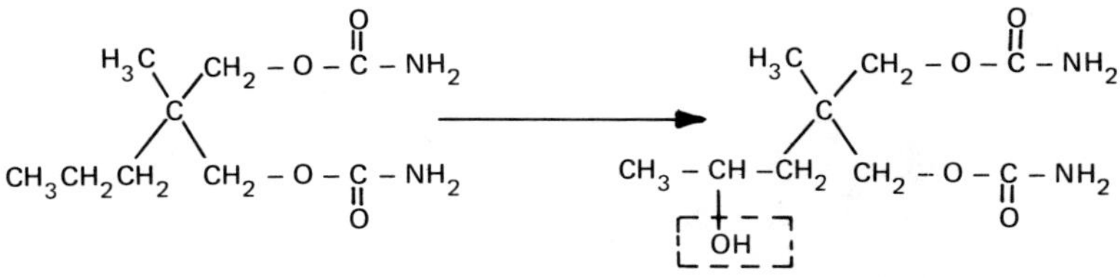

Meprobamate Hydroxymeprobamate

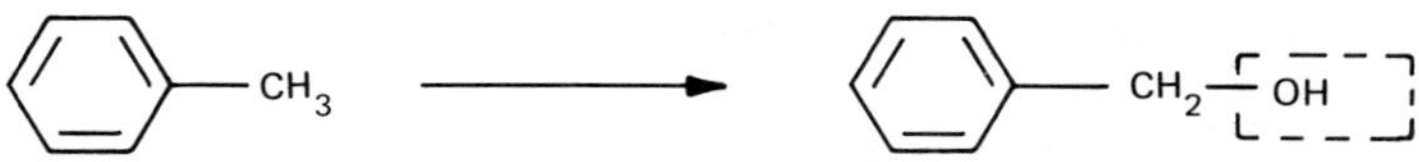

Toluene Benzyl alcohol

N-Dealkylation:

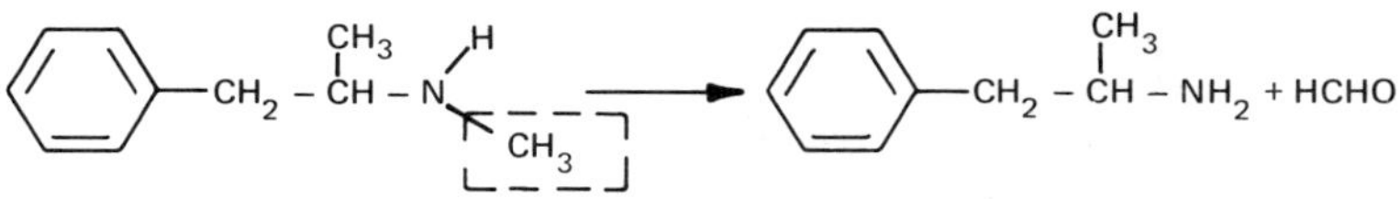

Methamphetamine Amphetamine Formaldehyde

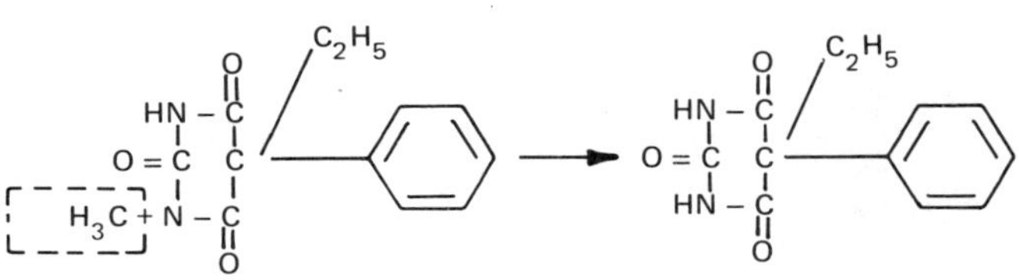

Mephobarbital Phenobarbital

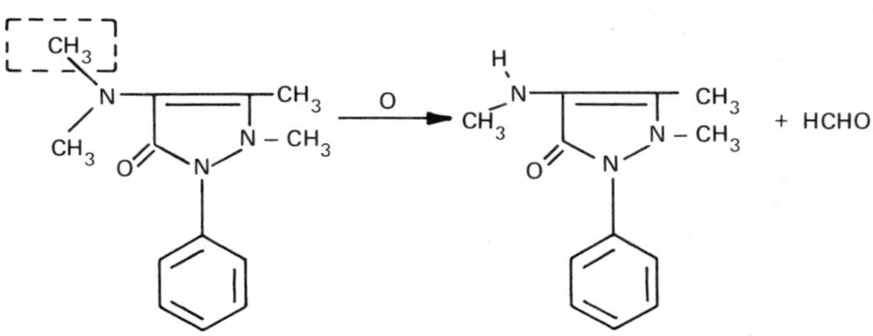

Aminopyrine 4-Methylaminoantipyrine

O-Dealkylation:

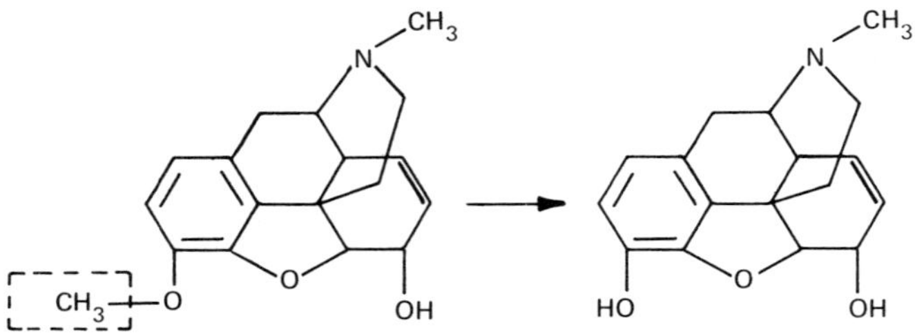

Codeine Morphine

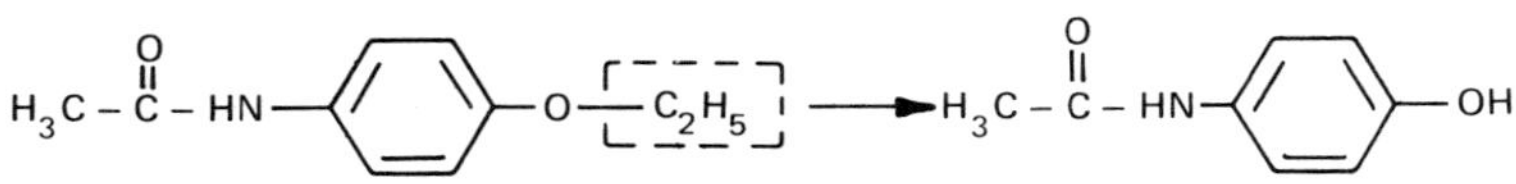

Phenacetin Acetaminophen (Tylenol)

S-Dealkylation:

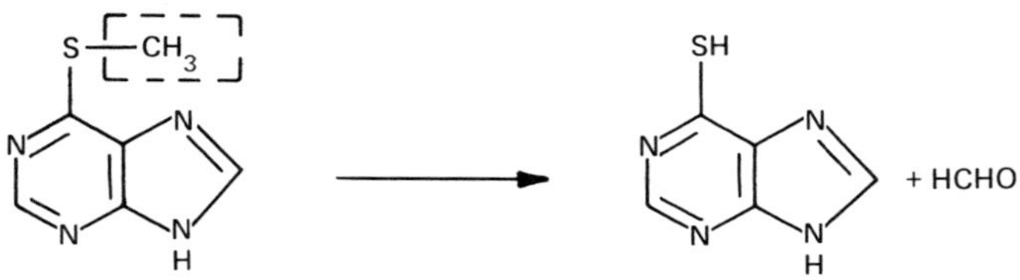

6-Methyl thiopurine 6-Mercaptopurine

Deamination:

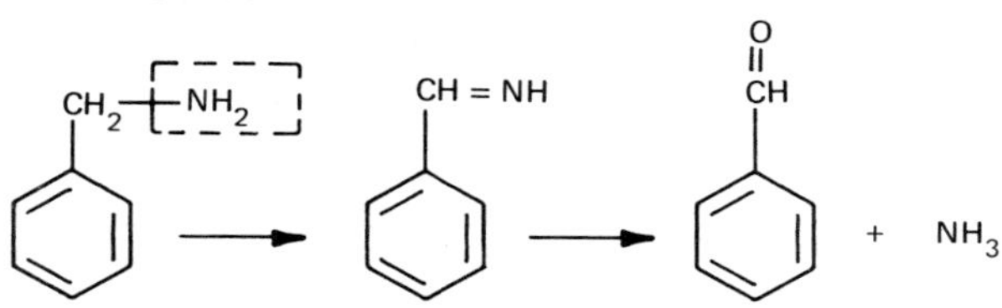

Benzylamine Benzylimine Benzaldehyde Ammonia

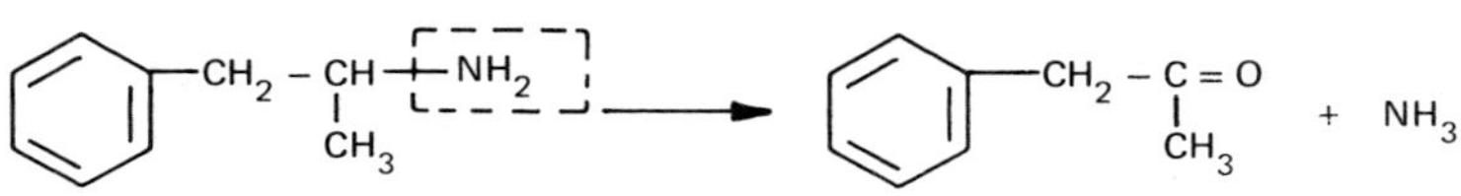

Amphetamine Phenylacetone

N-Oxidation:

$$(CH_3)_3{-}N \longrightarrow (CH_3)_3{-}N \rightarrow O$$

Trimethylamine — Trimethylamine oxide

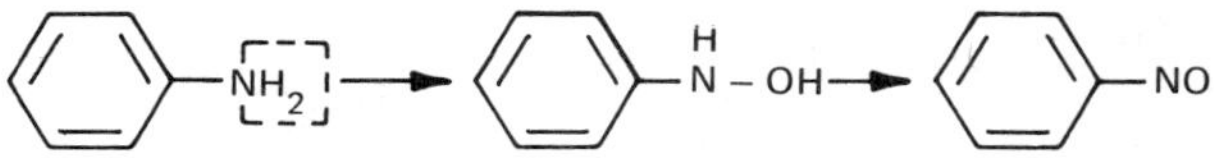

Aniline — Phenylhydroxylamine — Nitrosobenzene

Sulfoxidation:

Phenothiazine — Phenothiazine sulfoxide

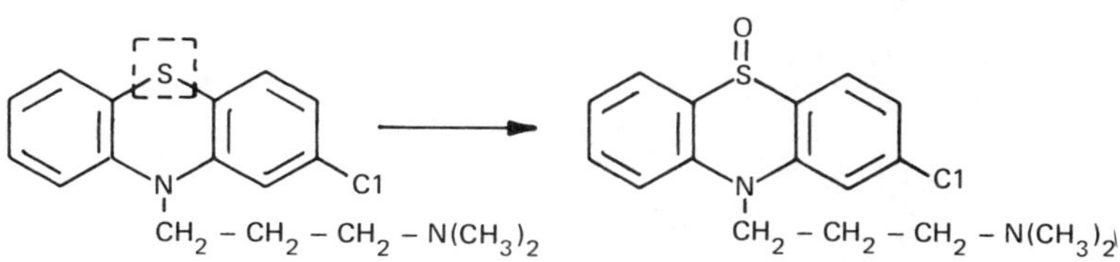

Chlorpromazine — Chlorpromazine sulfoxide

Desulfuration:

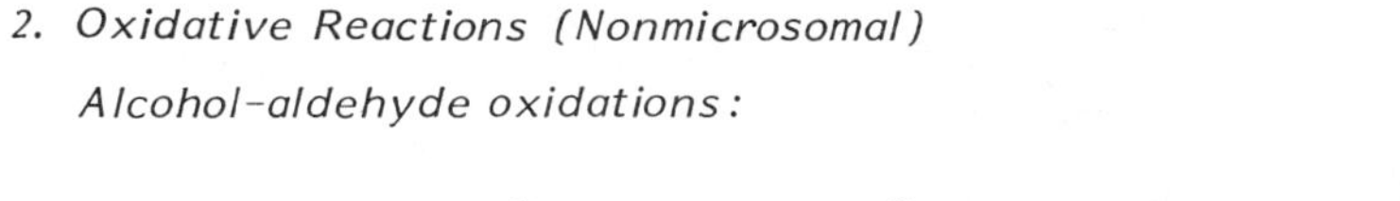

Parathion — Paraoxon

2. *Oxidative Reactions (Nonmicrosomal)*

Alcohol-aldehyde oxidations:

$$CH_3CH_2OH \longrightarrow CH_3-\overset{O}{\overset{\|}{C}}-H \longrightarrow CH_3-\overset{O}{\overset{\|}{C}}-OH$$

Ethanol — Acetaldehyde — Acetic acid

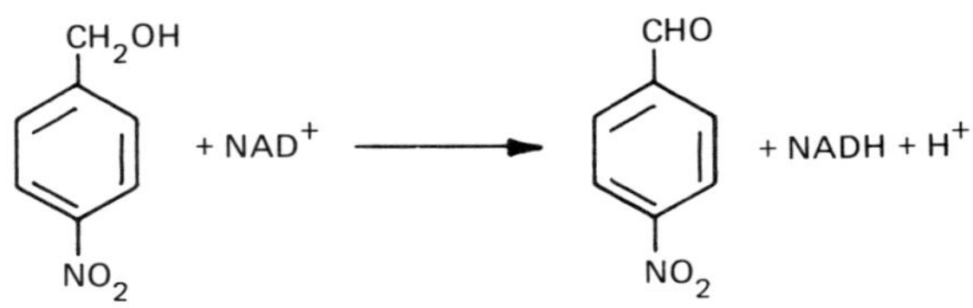

p-Nitrobenzyl alcohol p-Nitrobenzaldehyde

Ring aromatization:

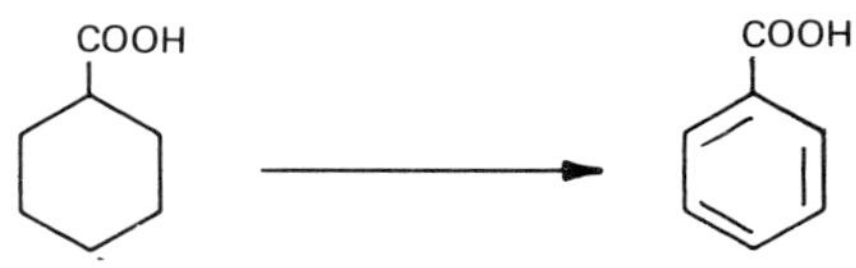

Hexahydrobenzoic acid Benzoic acid

3. *Reductions*

Aldehyde-ketone compounds:

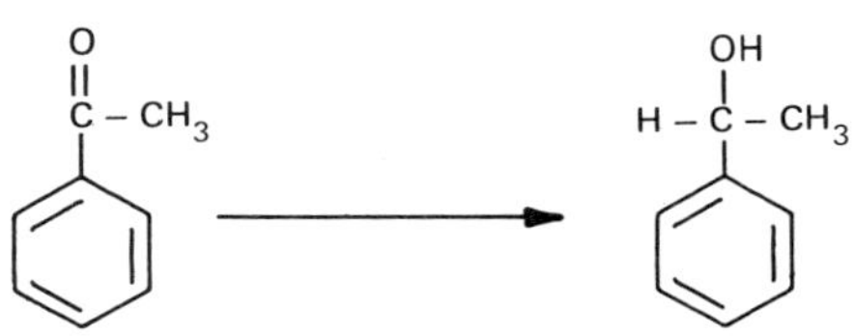

Acetophenone Phenylmethylcarbinol

Trichloroacetaldehyde monohydrate (chloral hydrate) 2,2,2-Trichloroethanol

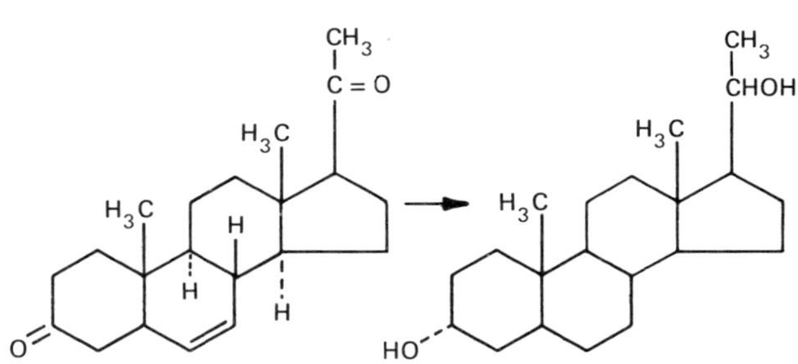

Pregn-4-ene-3,20-dione (Progesterone) 5-β-Pregnane-3α, 20α-diol (Pregnanediol)

Nitro compounds:

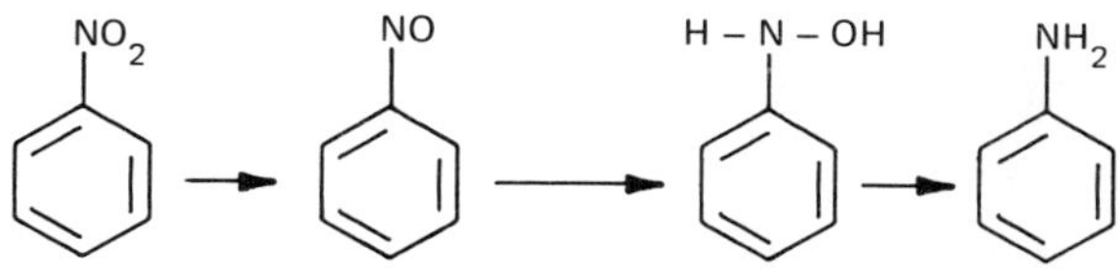

Nitrobenzene Nitrosobenzene Phenylhydroxylamine Aniline

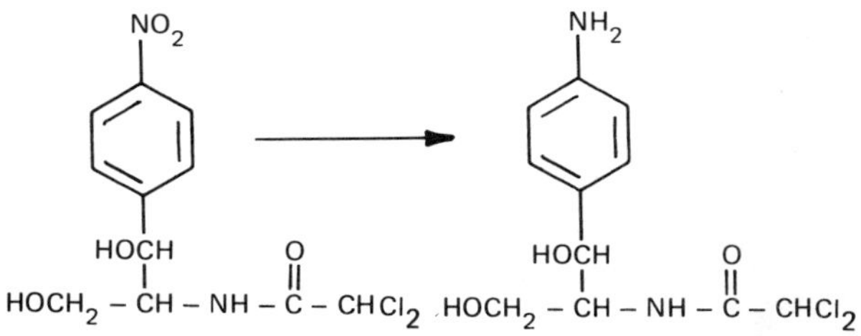

Chloramphenicol "Arylamine"

Azo compounds:

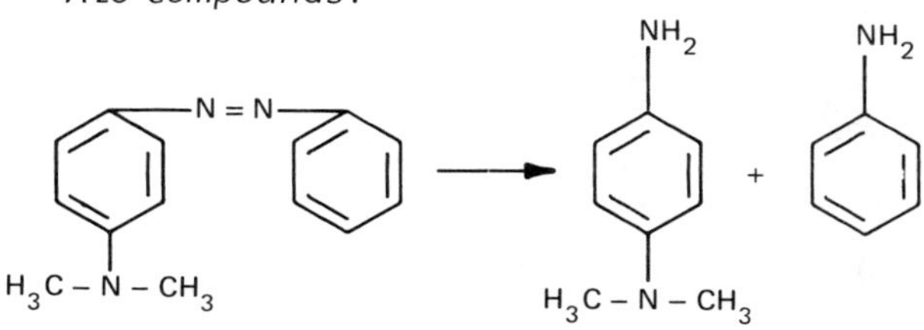

4-Dimethylaminoazobenzene 4-Dimethylaminoaniline Aniline

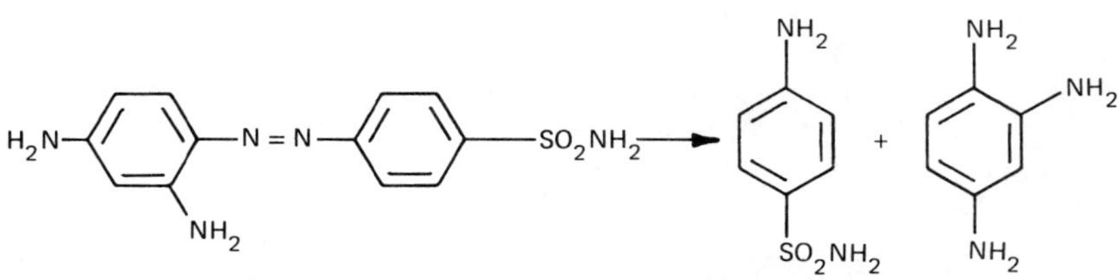

Prontosil Sulfanilamide 1,2,4-Triaminobenzene

4. *Hydrolysis*

Ester compounds:

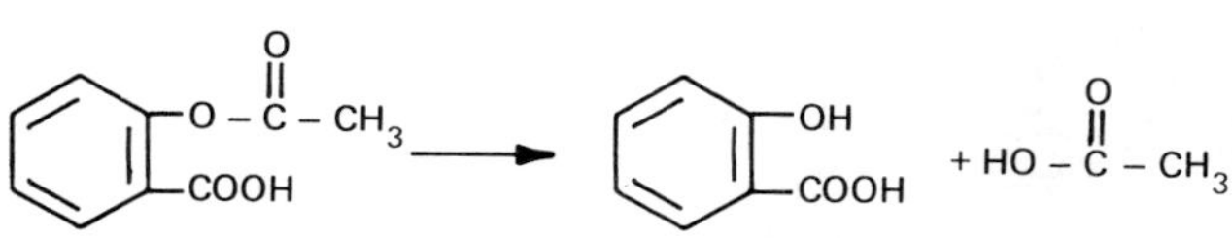

Aspirin Salicylic acid Acetic acid

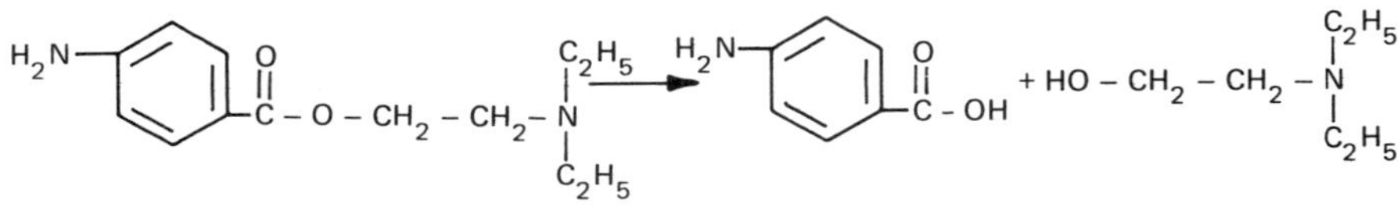

Procaine — *p*-Aminobenzoic acid (PABA) — Diethylaminoethanol

Amide compounds:

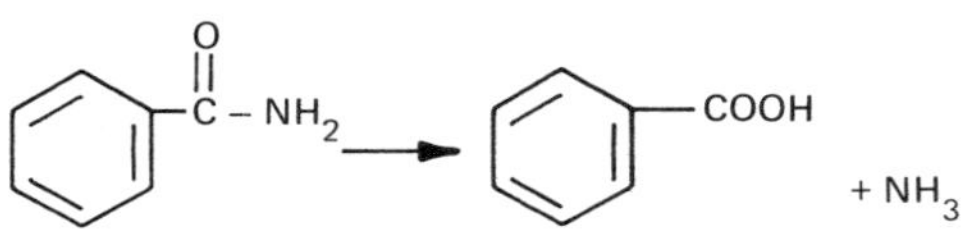

Benzamide — Benzoic acid

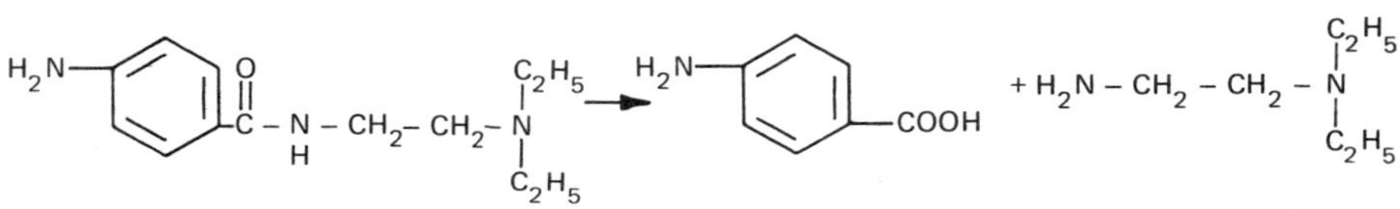

Procainamide — PABA — Diethylaminoethylamine

B. Synthetic Reactions

1. *Acylation*

To form an amide that can be metabolized, L-amino acids and alkyl and aryl amines are combined with organic acids. Initially, the mechanism of the acylation reactions involves the activation of the acid by the formation of a derivative with coenzyme A (CoA), this reaction involving ATP as noted below.

Whereas the endogenous acid is normally acetic acid, the endogenous amine may be either glutamine (in primates only), glycine (in many species), or ornithine or lysine (in birds). In both cases, the acid must be converted to an active form from which the acid is subsequently transferred to the amine with the aid of a transferase enzyme.

$$\text{R} - \text{COOH} + \text{ATP} \xrightarrow[\text{or thiokinase}]{\text{Acyl synthetase}} \text{R} - \text{CO} - \text{AMP} + \text{pyrophosphate}$$

$$\text{R} - \text{CO} - \text{AMP} + \text{CoA} - \text{SH} \xrightarrow[\text{kinase}]{\text{Acyl thio-}} \text{R} - \text{CO} - \text{S} - \text{CoA} + \text{AMP}$$

$$\text{R} - \text{CO} - \text{S} - \text{CoA} + \text{R'NH}_2 \xrightarrow[\text{acylase}]{\text{Trans-}} \text{R} - \text{CO} - \text{NHR'} + \text{CoA} - \text{SH}$$

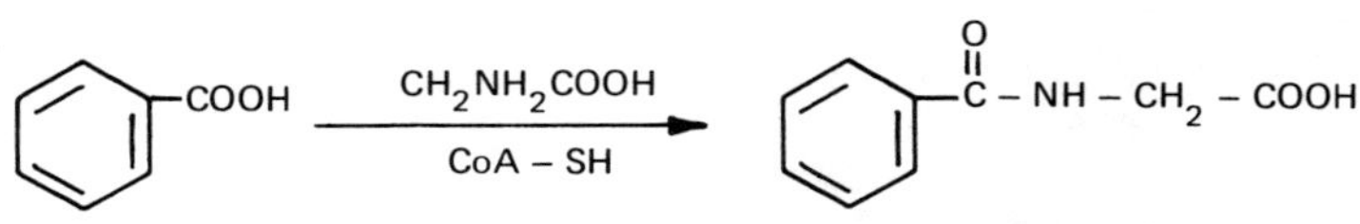

Benzoic acid Hippuric acid

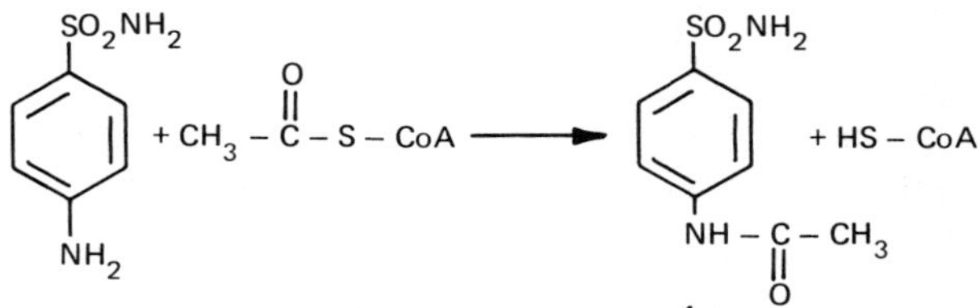

Sulfanilamide Acetyl CoA *N*[4]-Acetylsulfanilamide

In the overall series of reactions above leading to amide formation, RCOOH and $R'NH_2$ can be drugs that are to be conjugated.

2. O-, N-, and S-Methylations

In these reactions, the methyl group of S-adenosylmethionine is enzymatically transferred to the nitrogen of various amines (aliphatic or N-heterocyclics), the O of phenols, or the S of thiol groups. In the following reaction sequence, B may be either S, O, or NH.

$$\text{ATP} + \text{methionine} \xrightarrow[\text{transferase}]{\text{methionine adenosine}} \text{S-adenosylmethionine} + \text{pyrophosphate} + \text{phosphate}$$

$$\text{S-Adenosylmethionine} + \text{RBH} \xrightarrow[\text{transferase}]{\text{methyl}} \text{RBCH}_3 + \text{S-adenosylhomocysteine}$$

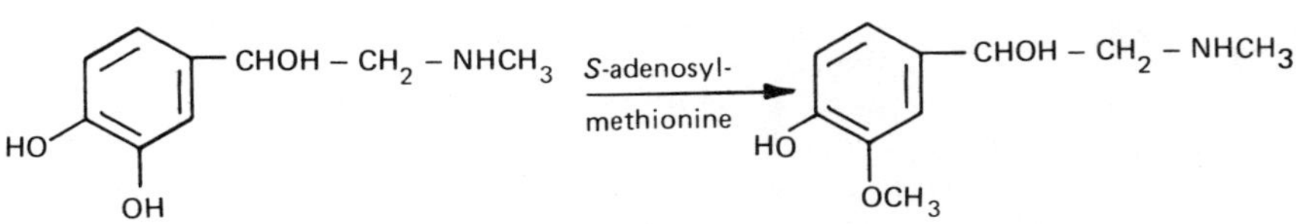

Epinephrine Metanephrine

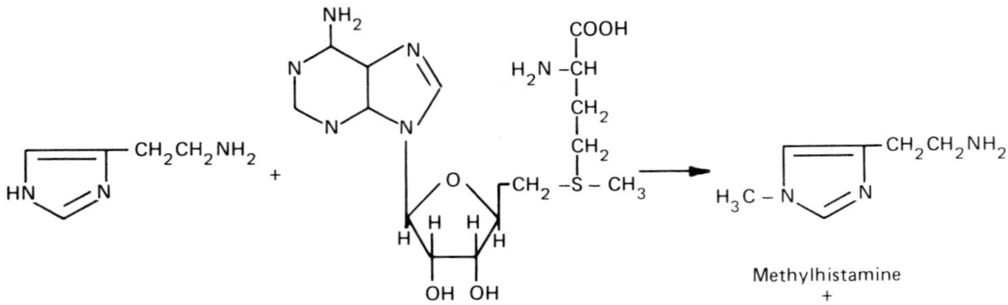

Histamine + *S*-Adenosylmethionine → Methylhistamine + *S*-Adenosylhomocysteine

Norepinephrine → Epinephrine

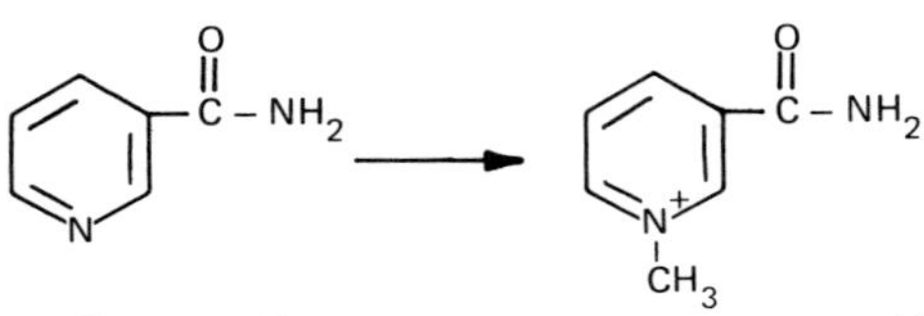

Nicotinamide → *N*-Methylnicotinamide

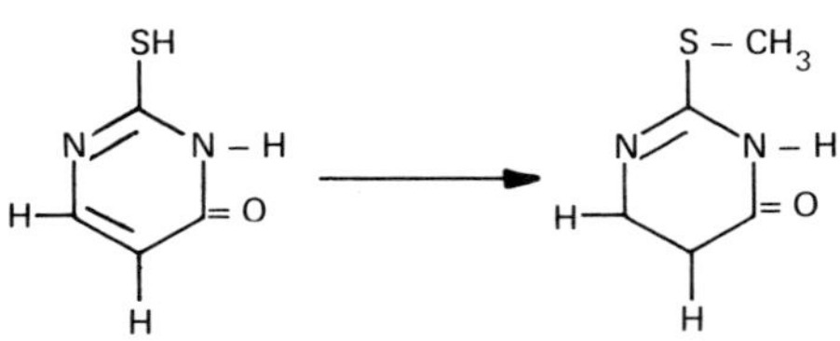

Thiouracil → *S*-Methylated compound

3. *Conjugation or Condensation Reactions*

Glucuronide formation: Glucuronide formation involves the condensation of glucuronic acid with the drug or its biotransformation product containing a reactive functional group; a group that may be either an amino ($-NH_2$), a hydroxyl ($-OH$), a sulfhydryl ($-SH$), or a carboxyl ($-COOH$). Uridine diphosphate-glucuronic acid (UDPGA) is synthesized catalytically by enzymes contained in the soluble fraction of the liver and serves as the donor of glucuronic acid to various acceptors. Common to all glucuronide derivatives is the free carboxyl group that is always present regardless of the conjugation process in question. The reactive hydroxyl group incorporated in the drug structure may be either alcohol (primary, secondary, tertiary), phenolic, enolic, or that due to hydroxyamines (=NOH); the carboxyl from either aromatic (carbocyclic, heterocyclic) or aliphatic (alkyl, aryl-alkyl) compounds; the amino from drugs that are aromatic, carboxyamide, sulfonamide, or heterocyclic in nature; and the sulfhydryl from either RSH or RSSH materials. Table 5 lists the various groups of exogenous compounds to which glucuronic acid can be transferred.

Table 5 Types of Groups in Foreign Compounds to Which Glucuronic Acid Can Be Transferred

Group	Structure	Example
Amino and imino		
Aromatic	ArCOOH	4,4'-Diaminodiphenylsulfone
Carbamate	$-OCONH_2$	Meprobamate
Heterocyclic	>NH	Sulfisoxazole
Sulfonamide	$-SO_2NH-$	Sulfadimethoxine
Carboxyl		
Aliphatic		
Primary	$-CH_2COOH$	Indomethacin
Secondary	>CHCOOH	Iodopanoic acid
Tertiary	≡CCOOH	Trimethylacetic acid
Aromatic	ArCOOH	Salicylic acid
Hydroxyl		
Alcoholic		
Primary	$-CH_2OH$	Trichloroethanol, chloramphenicol
Secondary	>CHOH	
Tertiary	≡COH	t-Butyl alcohol
Enolic	-CH=COH	4-Hydroxycoumarin
Hydroxylamine	-NOH	N-Hydroxy-N-2-fluorenylacetamide
Phenolic	ArOH	Morphine, p-acetamidophenol
Sulfhydryl		
Carbodithioic	-CSSH	
Thiol	-SH	Diethyldithiocarbamic acid

Source: Adapted from Smith and Williams [123], courtesy of Academic Press.

The general mechanisms involved are:

$$\text{Glucose-1-phosphate} + \text{UTP} \xrightarrow[\text{phorylase}]{\text{pyrophos-}} \underset{\text{(UDPG)}}{\text{UDP-glucose}} + \underset{\text{phate}}{\text{pyrophos-}}$$

$$\text{UDPG} + 2\text{NAD}^+ + \text{H}_2\text{O} \xrightarrow[\text{dehydrogenase}]{\text{UDPG-}} \underset{\text{(UDPGA)}}{\text{UDP-glucuronic acid}}$$

$$+ 2\text{NADH} + 2\text{H}^+$$

$$\text{UDPGA} + \text{RBH} \xrightarrow[\text{transferase}]{\text{glucuronyl}} \text{RB-glucuronic acid} + \text{UDP}$$

where B may be O, COO, NH, or S.

O-Ether glucuronides:

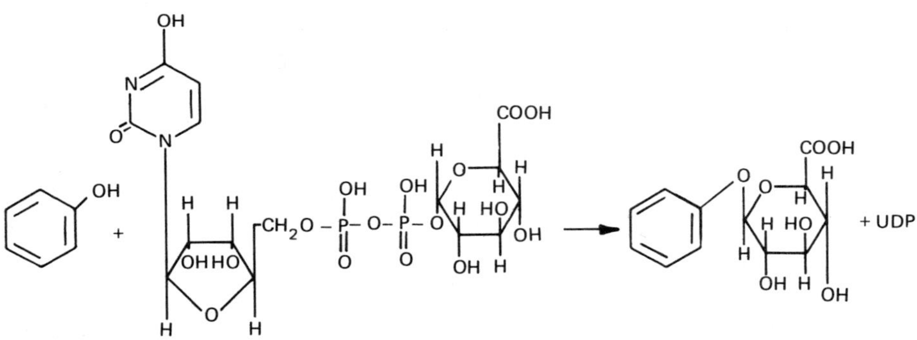

Phenol UDPGA Phenyl glucuronide

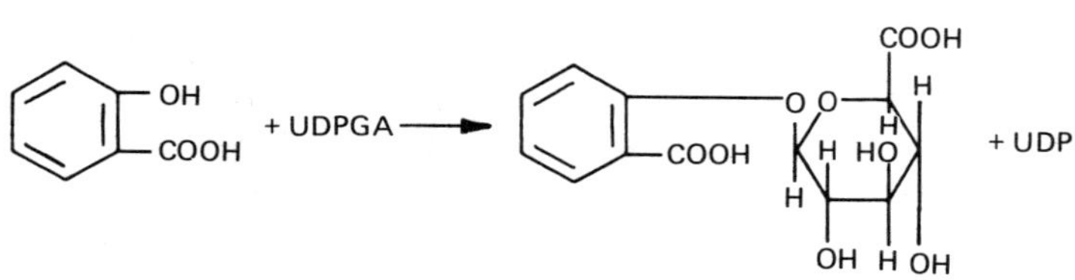

Salicylic acid *O*-Ether of salicylic acid

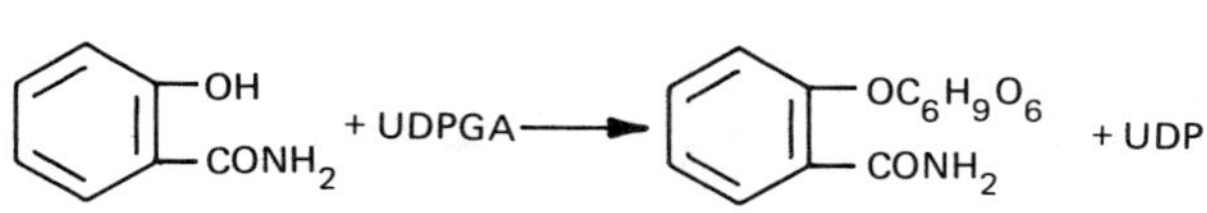

Salicylamide Salicylamide glucuronide

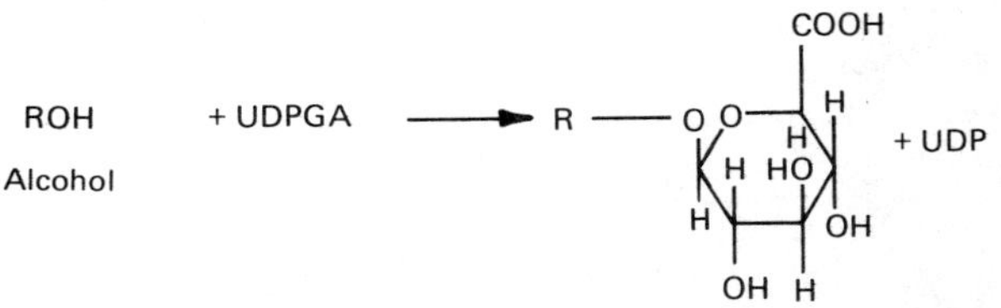

Alcohol

Ester glucuronides:

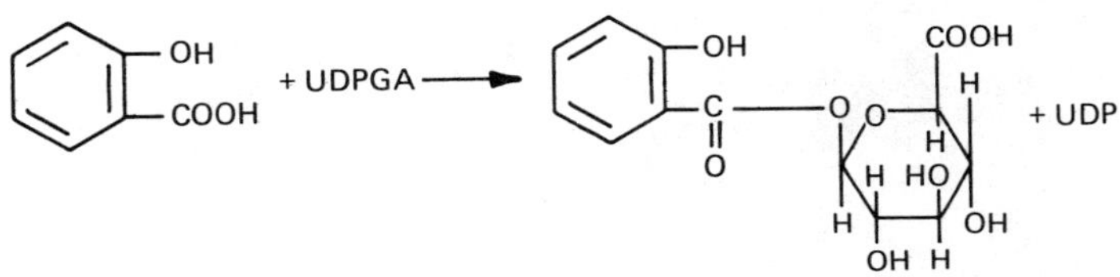

Salicylic acid

Glucuronide of salicylic acid

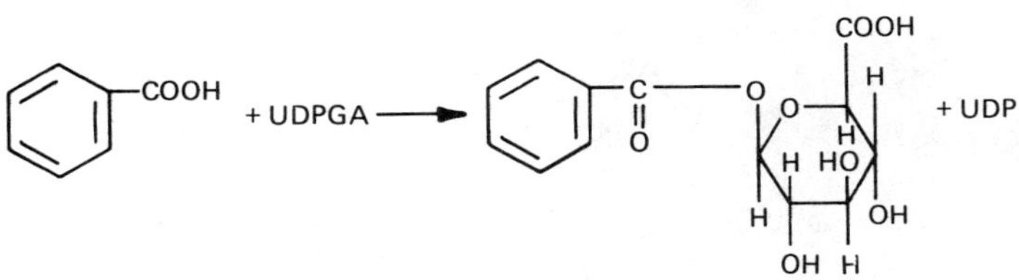

Benzoic acid

Benzoyl glucuronide

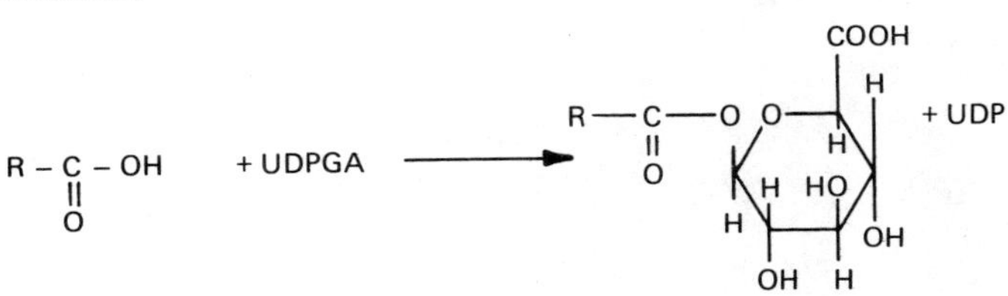

Carboxylic acid

Ester-*O*-glucuronide

N-Glucuronides:

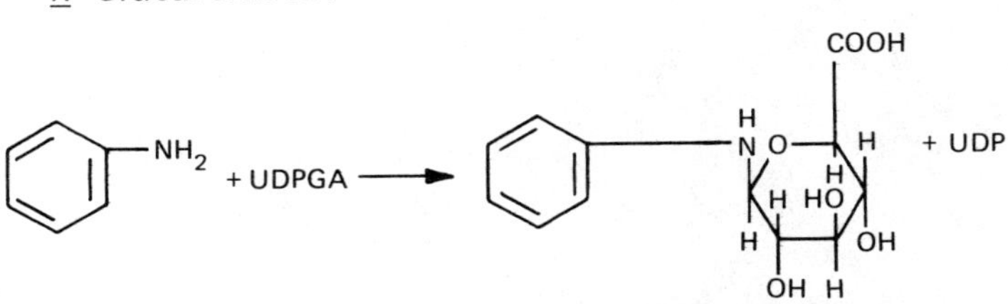

Aniline

Aniline glucuronide

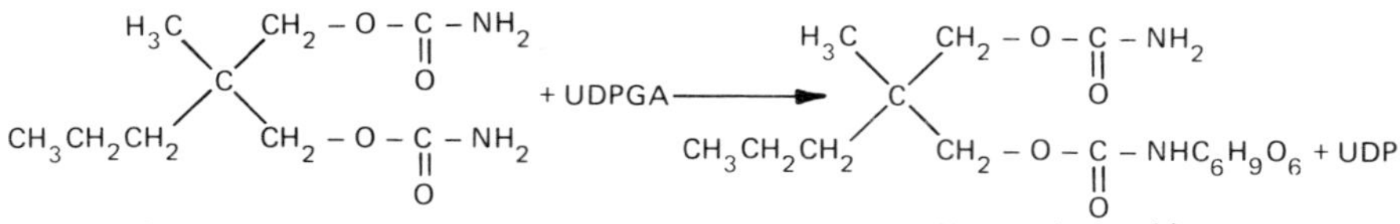

Meprobamate   Meprobamate-*N*-monoglucuronide

S-Glucuronides:

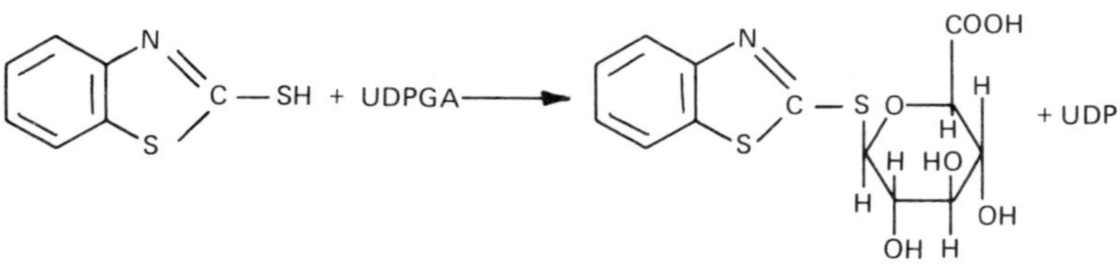

2-Mercaptobenzothiazole (2-MBT)   2-MBT-glucuronide

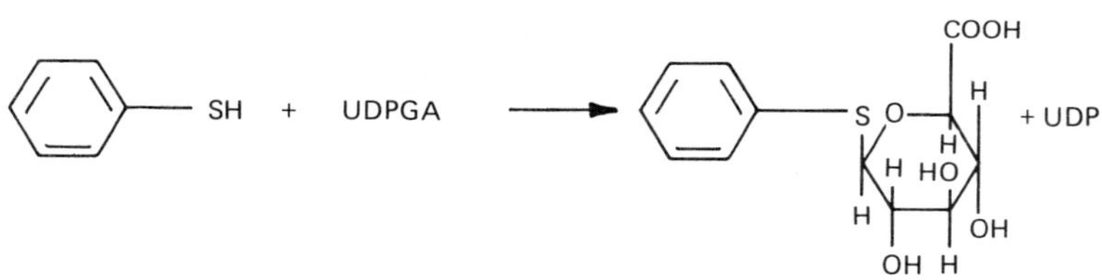

Thiophenol   Thiophenylglucuronide

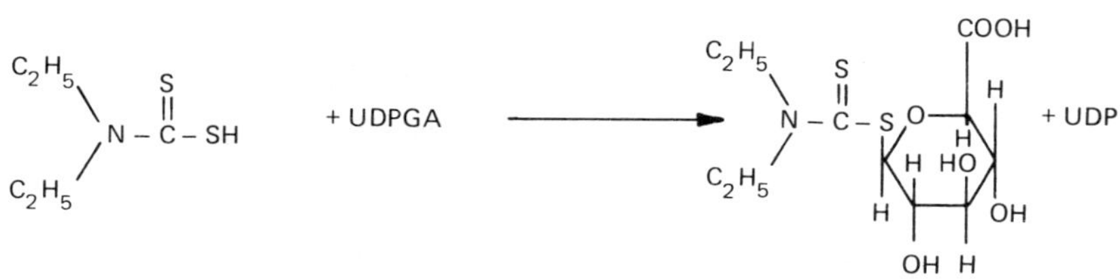

N,N-Diethyldithiocarbamic acid

Sulfate conjugation: Often called "ethereal sulfates," the sulfuric acid esters are synthesized by the reaction of an activated form of sulfate with either aliphatic hydroxyl and phenolic or certain amino groups. The inorganic sulfate is first converted to a high-energy nucleotide, adenosine-5'-phosphosulfate (APS); this is followed immediately by a further reaction with ATP, which yields 3'-phosphoadenosine-5'-phosphosulfate (PAPS). The sulfate group in PAPS is then transferred to the proper -OH or $-NH_2$ acceptor in in the presence of enzymes called sulfotransferases. These reactions, which occur in the soluble fraction of cells, are shown below.

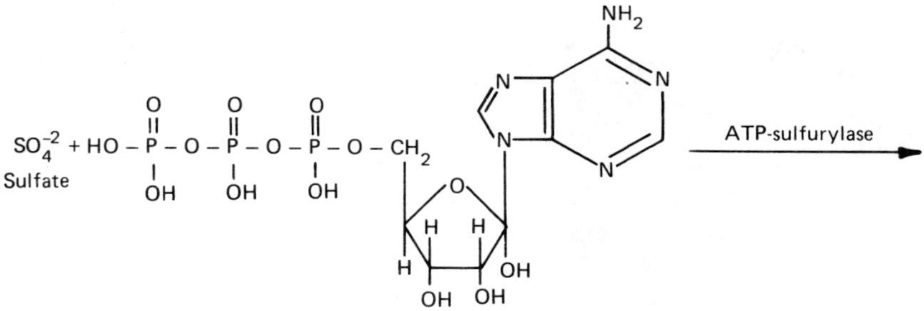

Adenosine triphosphate (ATP)

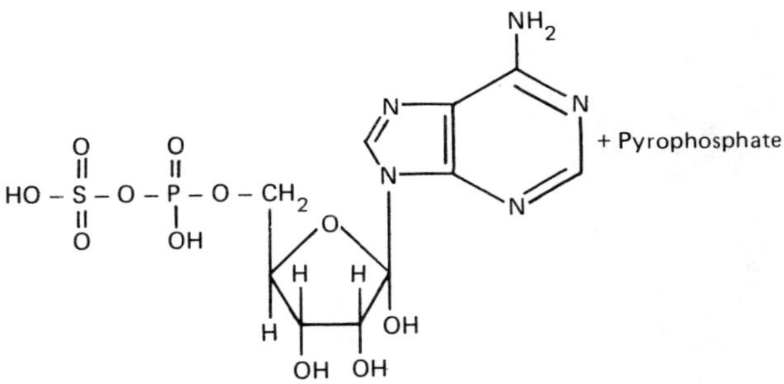

Adenosine-5′-phosphosulfate (APS)

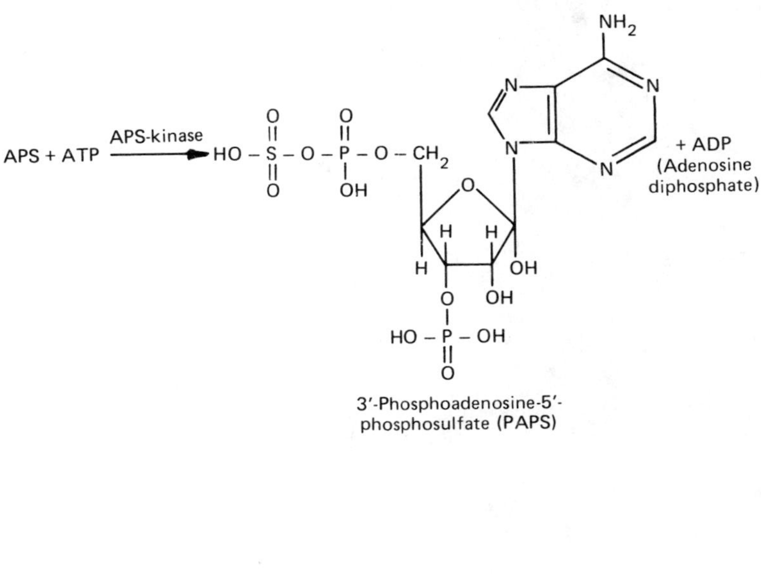

3′-Phosphoadenosine-5′-phosphosulfate (PAPS)

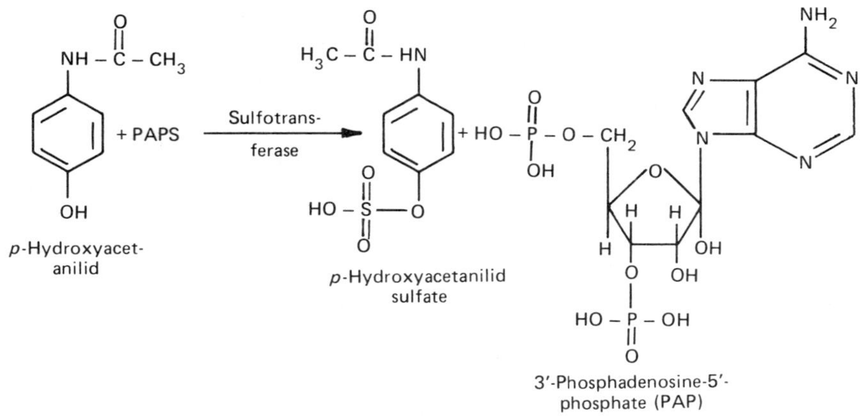

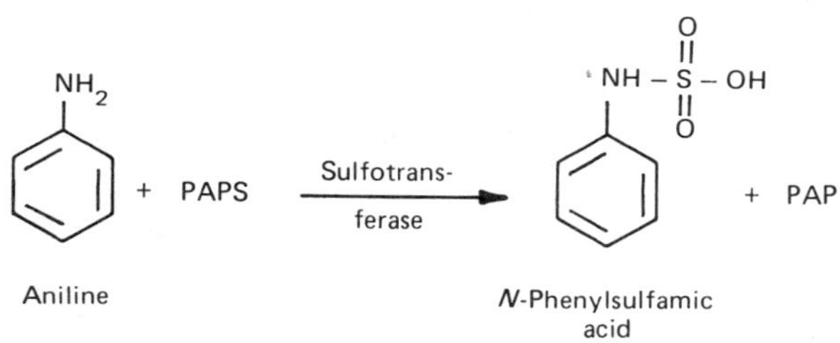

Amino acid conjugation: The condensation of a carboxylic acid-containing drug with an amino acid (glycine, glutamine, or ornithine) is preceded by the conversion of the acid into an active ester, S-acylcoenzyme A. By an appropriate enzyme the acyl group is transferred to the amino group of the amino acid. The overall series of reactions leading to the amide formation is as follows:

$$\mathrm{R\text{-}\underset{\underset{O}{\|}}{C}\text{-}OH + ATP \xrightarrow[\text{or synthesase}]{\text{acyl thiokinase}} R\text{-}\underset{\underset{O}{\|}}{C}\text{-}AMP + pyrophosphate}$$

$$\mathrm{R\text{-}\underset{\underset{O}{\|}}{C}\text{-}AMP + CoA\text{-}SH \xrightarrow[\text{kinase}]{\text{acyl thio-}} R\text{-}\underset{\underset{O}{\|}}{C}\text{-}S\text{-}CoA + AMP}$$

$$\mathrm{R\text{-}\underset{\underset{O}{\|}}{C}\text{-}S\text{-}CoA + H_2N\text{-}R'\text{-}COOH \xrightarrow[\text{acylase}]{\text{trans-}} R\text{-}\underset{\underset{O}{\|}}{C}\text{-}\overset{H}{N}\text{-}R'\text{-}COOH + CoA\text{-}SH}$$

where $H_2N\text{-}R'\text{-}COOH$ can be glycine, glutamine, or ornithine.

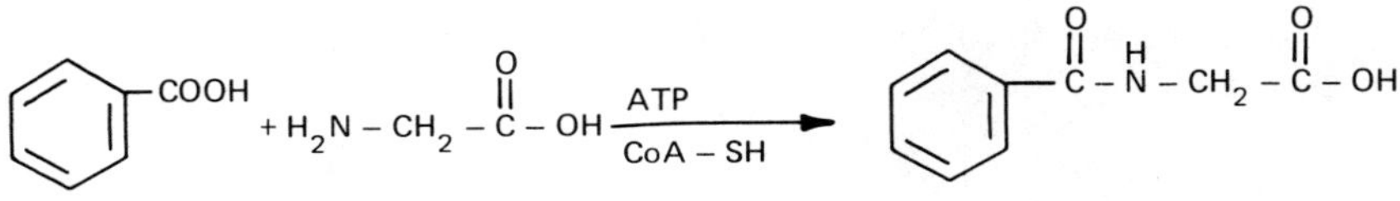

Benzoic acid Glycine Hippuric acid

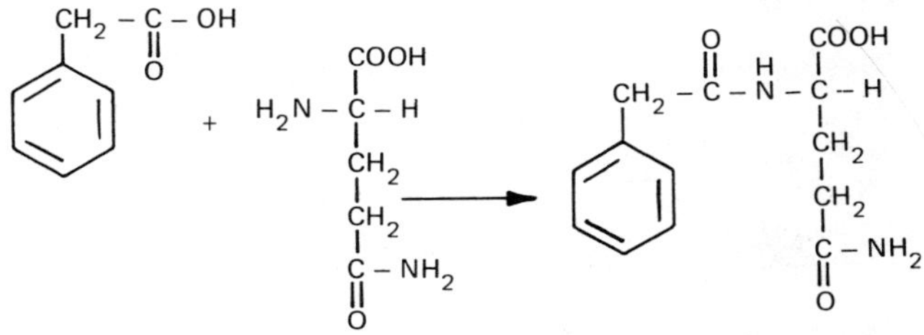

Phenylacetic acid Glutamine Phenylacetylglutamine

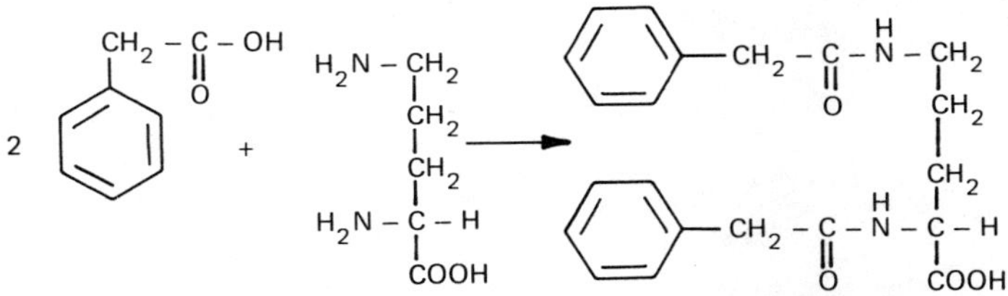

Phenylacetic acid Ornithine Diphenylacetylornithine

Synthesis of mercapturic acids: Drugs that form mercapturates frequently contain an active halogen or a nitro group. It has been postulated that the reaction proceeds in the following manner:

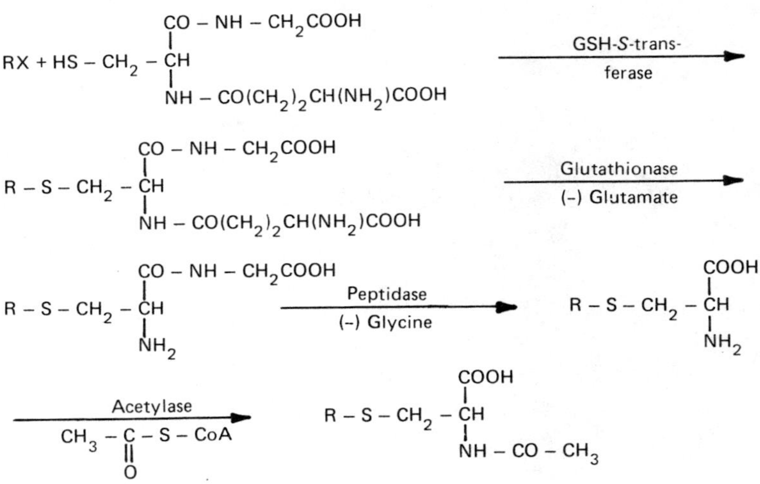

Several compound types that form mercapturic acids in vivo are shown below:

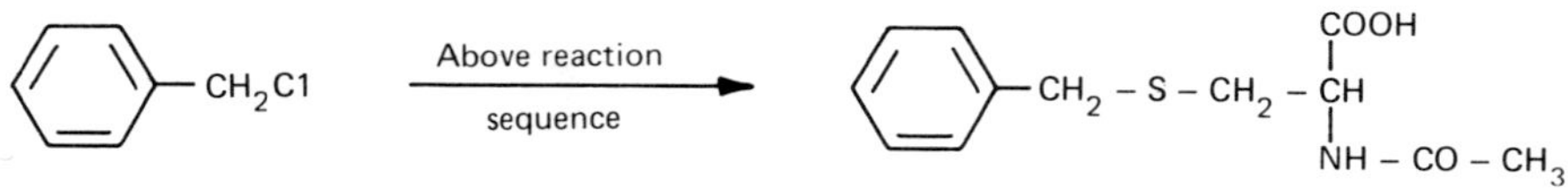

Benzyl chloride   Benzyl mercapturate

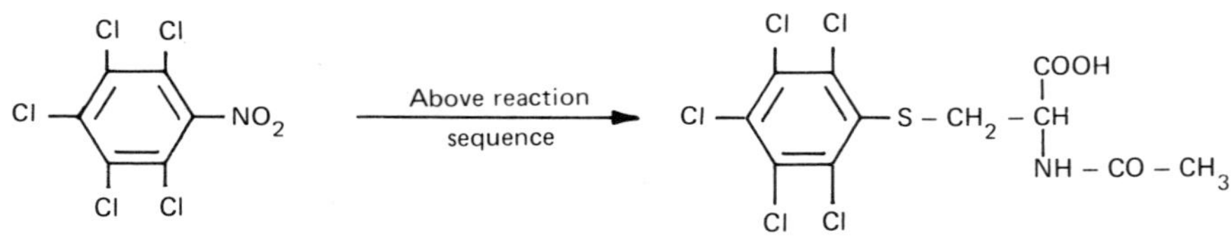

Pentachloronitrobenzene   Pentachlorobenzomercapturate

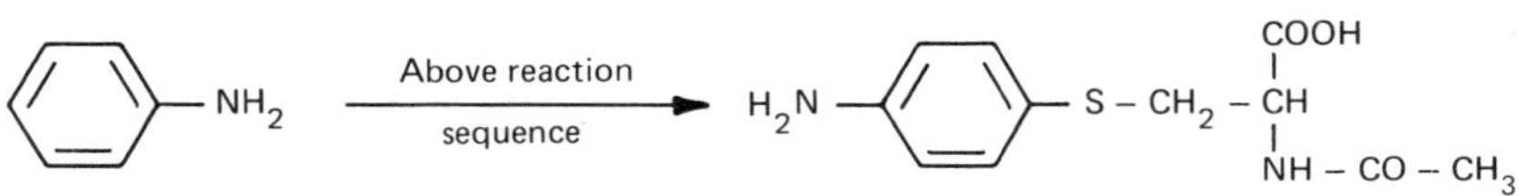

Aniline   *p*-Aminophenylmercapturate

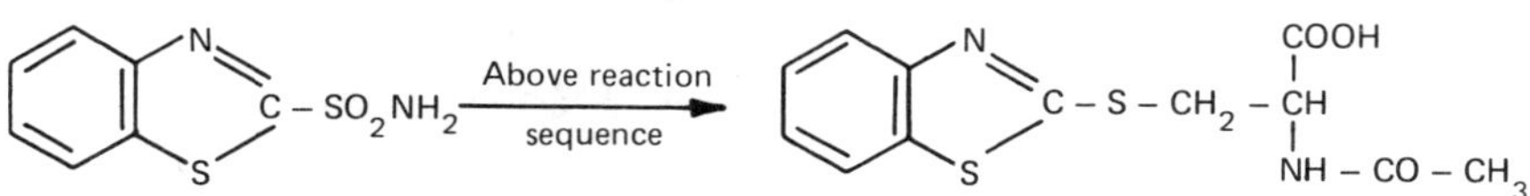

Benzothiazole sulfonamide   Benzothiazole mercapturate

Certain α-halogenated alkylbenzenes may conjugate directly with acetylcysteine and are excreted mainly as S-benzylmercapturic acids. For example, 2,4-dichloronitrobenzene and benzyl chloride form 3-chloro-4-nitrophenylmercapturic and S-benzylmercapturic acids, respectively.

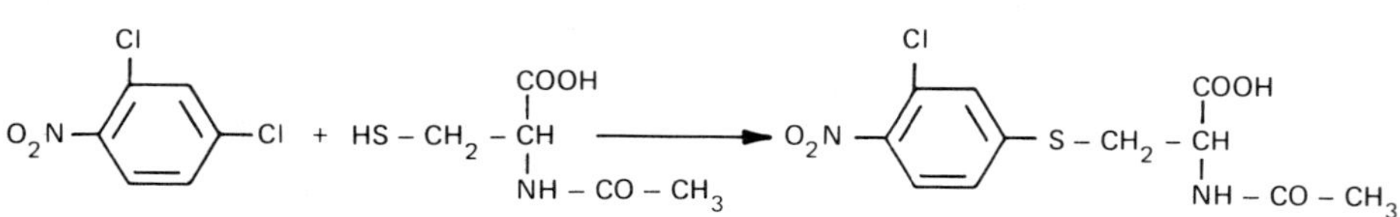

2,4-Dichloronitrobenzene   Acetylcysteine   3-Chloro-4-nitrophenylmercapturic acid

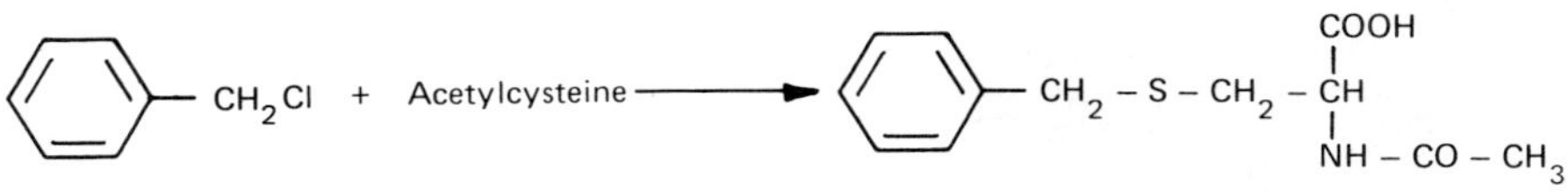

Benzyl chloride

S-Benzylmercapturic acid

Benzene, naphthalene, and anthracene can also be transformed into mercapturates where the aromatic is initially transformed to an intermediate epoxide product. In turn, this epoxide reacts with glutathione to form a "premercapturate," which in the presence of dilute acid, becomes the corresponding mercapturate by the loss of a water molecule. Using naphthalene as an example, the reactions required to obtain the 1-naphthylmercapturic acid product are as follows [124]:

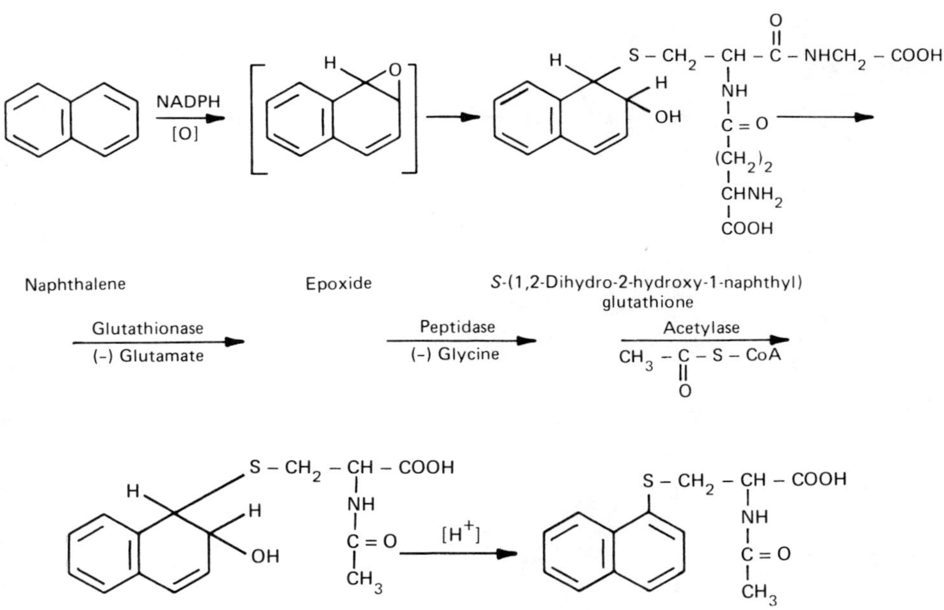

Naphthalene

Epoxide

S-(1,2-Dihydro-2-hydroxy-1-naphthyl) glutathione

N-Acetyl-*S*-(1,2-dihydro-2-hydroxy-1-naphthyl) cysteine "Premercapturate"

1-Naphthylmercapturic acid

4

Elimination or Excretion of Drugs

Drug action in the body can be terminated by either of two processes: (a) biotransformation or (b) excretion via the lung, the kidney, or the liver. As noted by Gourley [70] in Fig. 5, unbound or free drugs are eliminated from the body either directly as the unchanged molecule or indirectly as the metabolic product(s) of the parent compound, or by both routes. Of the routes of elimination cited above, the major mechanisms are hepatic metabolism and renal excretion. Generally, the more polar compounds are excreted unchanged, whereas the less polar, lipid-soluble drugs are not readily eliminated until they have been biotransformed to more polar, less lipid-soluble materials. Nevertheless, the mechanisms of passage across membranes involved in excretion are identical for both the unchanged parent compounds and their metabolites.

When the rate of absorption exceeds the rate of elimination, drug concentration in blood and tissues is increased. When these rate processes are reversed, elimination will be predominant.

I. PULMONARY DISPOSITION OF DRUGS

Pulmonary elimination of drugs which are excreted largely in their original unchanged form is of interest only in relation to the inhalation of anesthetics (ethylene, cyclopropane, nitrous oxide, halothane, chloroform, diethyl ether, etc.). As pointed out by

Wollman and Dripps [125], "the depth of general anesthesia varies directly with the partial pressure (tension) of anesthetic agent in the brain and the rates of induction and recovery depend upon the rate of change of tension in this tissue." The factors that determine the anesthetic gas tension in the arterial blood and in the brain are essentially four in number: (a) partial pressure of the anesthetic in inspired gas, (b) pulmonary ventilation bringing gas into the lungs, (c) transfer of the anesthetic gas from the alveoli to the blood flowing through the lungs, and (d) loss of the anesthetic agent from the arterial blood to other body tissues. In turn, the transfer of anesthetic gases from alveoli to blood is dependent on three factors in the absence of ventilation-perfusion disturbances: (a) solubility of the anesthetic in blood, (b) rate of cardiac output (pulmonary gas flow), and (c) the magnitude of the differences in gas partial pressures (tensions) between alveoli and pulmonary venous blood. The rate of gas uptake in blood and tissues can be mathematically expressed by its partition coefficient which is numerically equivalent to the Ostwald solubility coefficient (the ratio of the anesthetic's concentration in both phases). Hence gas uptake will be greater for those agents having higher solubilities, blood-to-gas ratios, or partition coefficients. However, when gases are administered at rather low concentrations, the rate at which the agent's alveolar concentration rises toward that inspired is inversely proportional to solubility. On the other hand, an increase in pulmonary gas flow definitely affects the rate at which the anesthetic gas passes from the alveolar gases into the arterial blood, as indicated by the initial portion of the arterial tension curve. The delay in the rate of rise of the alveolar concentration is simply a function of a greater blood volume exposed to the gas within the lung. As for the magnitude of the alveolar-venous difference in gas partial pressures, it is greatest at the time of least saturation but steadily diminishes as an equilibrium is achieved between the alveolar partial pressure and that of all body tissues.

Of primary importance in the determination of the rate of elimination of an anesthetic or drug via the lungs is the blood-to-air partition coefficient [70]. For this reason, anesthetics with low blood-to-air partition coefficients will foster rapid rates of recovery after anesthetic administration is terminated. This important aspect of the blood-to-air partition coefficient in determining the anesthetic's rate of elimination from the body via the lungs has been discussed by Butler [126, 127].

Although the transfer of anesthetic gases in the body is also influenced by their metabolism (i.e., their partial or complete removal by the liver through which 25% of the cardiac output flows, Wollman and Dripps [125] summarize the uptake, distribution, and elimination in inhalation anesthetics as follows:

> In summary, during the administration of an anesthetic, its tension in blood rises toward that in the inspired gas, at first rapidly, then more slowly. Tissue tensions then increase, approaching the arterial tension. The partial pressure increases most rapidly in tissues with high rates of blood flow, and lags considerably in areas where blood flow is slower. Before arterial gas tension has reached the inspired tension, the administration is usually over. As the gas is eliminated, its partial pressure in the arterial blood declines first, followed by that in the tissues. The major factors that affect rate of elimination of the anesthetics are the same as those that are important in the uptake phase.

II. RENAL DISPOSITION OF DRUGS

Renal excretion of drugs in the urine may involve any or all of the following processes [9,63,70,128]: (a) glomerular filtration, (b) active tubular secretion, and (c) passive tubular reabsorption. The filtration rate and the magnitude of plasma protein binding are the primary factors that determine the quantity of drug entering the lumen by filtration. In the glomerulus, passage of nearly all solutes is permitted since glomerular filtration is essentially a nonselective process. However, higher molecular weight materials such as free proteins are excluded from the glomerular filtrate which passes in sequence through the lumens of the proximal tubule, loop of Henle, and the distal tubule before entering the collection duct [70,128]. During its passage, organic solutes are concentrated since 99% of the water filtered in the glomerulus is reabsorbed. In the proximal renal tubule, strong organic bases and acids enter the glomerular filtrate by active, carrier-mediated tubular secretion. Furthermore, depending on the concentrations of the drug and the pH on the two sides of the tubular cells, the non-ionized species of weak acids and bases can undergo reabsorption and excretion by passive diffusion, a bidirectional mechanism whereby the drugs may diffuse across tubular cells. In general, drug excretion is pH dependent if the pK_a of the drug is in the range of 7.15-10.5 for weak bases and 3.0-7.5 for weak acids [70]. Hence, as pointed out by Csaky [12], a drug excretion via the kidney can be (a) filtered in the glomeruli but neither reabsorbed nor excreted in the tubules, (b) filtered and partially reabsorbed, or (c) filtered and secreted in the tubules. To determine which of these three possibilities is followed by a given drug, clearance values are employed. This approach has also been used to describe elimination by the liver, bile, lungs, a metabolic route, or for that matter, any means by which the body excretes or transforms the drug [9].

Renal or plasma clearance, C_r, may be defined as the plasma volume in milliliters which is cleared of a given substance by the kidneys in 1 min. It can be determined from

$$C_r = \frac{C_u V_{ue}}{C_p} \tag{101}$$

where

C_u = concentration of the drug in the urine, mg/ml
V_{ue} = volume of urine excreted in 1 min, ml/min
C_p = concentration of the drug in the plasma, mg/ml

As pointed out by Notari [9], this is equivalent to defining C_r as the amount excreted in 1 min/amount contained in 1 ml of plasma.

If a substance is not bound to plasma proteins (i.e., it is presumed to be freely filtered in the glomeruli), is neither reabsorbed nor secreted, and is nontoxic, its clearance measures the volume of plasma filtered through the glomeruli in 1 min. Hence it will be filtered in amounts equal to

$$C_f = C_p \times \text{GFR} \tag{102}$$

where GFR is the rate of glomerular filtration expressed in units of ml/min. On the other hand, since the volume of the renal plasma flow (RPF) is filtered into tubular lumens, the filtration fraction (FF) is obtained by

$$\text{FF} = \frac{\text{GFR}}{\text{RPF}} \tag{103}$$

The polysaccharide inulin, mannitol, and sodium thiosulfate are completely filtered by the glomeruli and excreted in the urine. Since none of these materials is secreted or reabsorbed by the tubule, the clearance of each is therefore a measure of GFR; their clearance value is equal to approximately 125-130 ml.

On the other hand, low threshold substances such as urea, uric acid, certain phosphates, and sulfates are filtered by the glomeruli and passively reabsorbed in the tubules. Normally, the clearance value for urea is slightly less than 75. The urea clearance value is substantially less than that of inulin, which indicates that some of the filtered urea is subsequently passively reabsorbed by the tubules [9].

High threshold substances, such as glucose, ascorbic acid, Na, Ca, Mg, P as PO_4^{3-}, Cl, and S as SO_4^{2-}, are normally completely reabsorbed by active tubular reabsorpion.

Renal plasma flow can be determined by p-aminohippurate (PAH) which is filtered at the glomeruli and secreted by the tubules. At low concentration, it is completely removed from the plasma during one pass through the kidneys. Hence the amount of PAH in the urine is a measure of the volume of plasma cleared of PAH in a unit of time or is equivalent to approximately 650-700 ml.

By relating the clearance values of drugs to the glomerular filtration rate, it therefore becomes possible to compare the excretion rates of various drugs.

$$\text{Clearance ratio} = \frac{\text{plasma clearance (ml/min)}}{\text{GFR (ml/min)}} \qquad (104)$$

As pointed out by Csaky [12], more accurate information can be obtained concerning the rate of excretion of drugs by using as C_p the concentration of the drug not bound to plasma proteins. In this manner, the excretion ratio is obtained. By definition, for those drugs or substances that are filtered but neither reabsorbed nor secreted, the excretion ratio is 1.0 (as previously noted, these are substances used to measure GFR because plasma clearance = GFR). Values less than 1.0 indicate that the drug is filtered and partially reabsorbed; values greater than 1.0 show that the drug is excreted by the tubules.

Expressed in terms of clearance values, a value of 130 ml would indicate glomerular filtration, a value greater than 130 ml would indicate both filtration and secretion, and a value less than 130 ml would then indicate passive reabsorption.

III. HEPATIC DISPOSITION OF DRUGS

The biliary tract represents another principal route by which many metabolites of drugs formed in the liver are excreted from the body. Excreted into the intestine via the bile in the form of conjugates, these still might be reabsorbed unchanged into the plasma or excreted unchanged into the feces. As noted by Williams et al. [129], it is also possible to hydrolyze some conjugates by enzymes present in the intestine or in intestinal bacteria before subsequent reabsorption or excretion. In contrast to drugs of less than 150 molecular weight that are mainly excreted in the urine, the larger drug structures are normally excreted into the intestine via the bile provided, however, that they can be metabolized and conjugated. Williams et al. [129] and Cox and Wright [130] have reported from their investigations that it appears that a molecular weight greater than 300 and the presence of polar groups facilitate biliary excretion. Conjugation is also regarded as important in the biliary excretion of drugs and of special interest is the conjugation of the drug or its metabolites with glucuronic acid. However, there are some exceptions to these generalizations; for example, the correlation between polarity and excretion reported by Haddock et al. [131] and Meyer-Brunot and Keberle [132].

In his discussion of the passage of drugs across body membranes, Gourley [70] writes:

> The excretion in bile of large, lipid insoluble molecules indicates that the membrane between bile and hepatic sinusoidal blood is highly porous. This membrane permits the penetration of virtually all molecules and ions whose size is less than that of protein molecules. The hepatic parenchymal cell is also permeable to polar molecules to a limited extent but very permeable to lipid-soluble molecules. In addition, specialized transport processes apparently are available for concentrating acids and bases from blood into bile. For some drugs, the transport mechanism is saturable, and various drugs compete with one another for transport. The similarity between these transport mechanisms and those for acids and bases in the kidney is obvious.

In addition to pulmonary, renal, and biliary tract excretions, there are other routes for drug disposition: by mammary, sweat, salivary, and the lachrymal and reproductive tract secretions. The extrarenal excretion of drugs and chemicals is a subject that has been excellently discussed and reviewed by Stowe and Plaa [133] in 1968.

IV. KINETICS OF DRUG ELIMINATION

Depending on the mechanism of elimination (a term used to decribe all processes that reduce the effective drug concentration in body fluids), the course of the elimination or disappearance of a drug may be either zero-order (constant rate) or first-order (exponential) elimination. As in absorption, first-order elimination is the general rule, but if the elimination mechanism becomes saturated (as is the case with constant intravenous injections or infusions), the elimination will be zero-order when drug concentrations exceed the saturation level. When the concentration falls below the saturation level, the elimination rate will decrease with decreasing concentration.

A. First-Order Kinetics

1. Single-Compartment Model

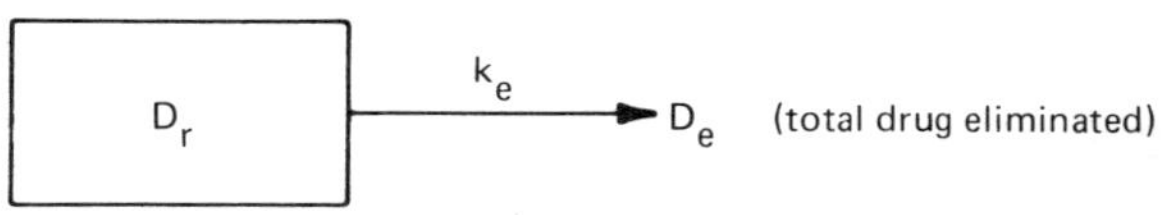

As previously noted, first-order elimination mechanisms are to be expected as a general rule; this being a situation where a constant fraction of the therapeutic agent disappears in each equal interval of time.

As pointed out by Goldstein et al. [1], after the distribution phase as been completed, the plasma concentration of drug (C_p) and the total amount of drug in the body ($D_r = Q = C_pV_d$ at time t, where V_d is the apparent volume of distribution) decline exponentially, following first-order kinetics. If D_r represents the total amount of drug in the body at time t, D_0 the quantity of drug present at t = 0, and k_e the elimination rate constant, then

$$D_r = D_0 e^{-k_e t} \qquad (105)$$

or

$$\log D_r = \log D_0 - \frac{k_e t}{2.30} \qquad (106)$$

If D_0 (equivalent to the dose or amount of drug administered) and k_e are known, it is possible to calculate at any time the amount of drug remaining in the body. If the logarithm of D_r is plotted against time, one can determine the k_e value since it is determined from the slope of the semilogarithmic plot.

Using a single-compartment open model to represent the body and to describe the kinetics of drug levels after single doses where both the absorption and elimination of drugs follow first-order kinetics, the serum concentration of drug (the amount absorbed into the circulation) for the cases where $k_a \neq k_e$ and $k_a = k_e$ can be derived by Eqs. (42) and (43), respectively.

Golman [134] used such a model to determine the absorption of metrizamide into the blood:

$$C_p = \frac{FD_0 k_a}{V_d(k_a - k_e)}\left[e^{-k_e(t-t_0)} - e^{-k_a(t-t_0)}\right] \quad (107)$$

where

C_p = drug concentration in serum
D_0 = dose or quantity of drug present at $t = 0$
F = fraction of D_0 absorbed
k_a = first-order rate constant for absorption
k_e = first-order rate constant for elimination or excretion
t = time after injection
t_0 = lag time
V_d = apparent volume of distribution

2. Two-compartment Open Model

Nayak et al. [135] used the two-compartment model with oral absorption shown schematically below to describe methaqualone pharmacokinetics after single- and multiple-dose administration in humans.

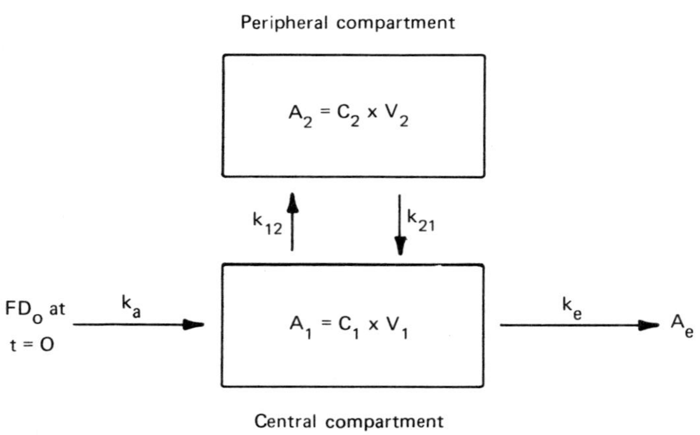

In this model, k_a and k_e are first-order rate constants for absorption and elimination; k_{12} and k_{21} are first-order rate constants for drug transfer between the central and peripheral compartments; A_1, A_2 and C_1, C_2 are the amounts and concentration of drug in the serum of the central compartment at time t and in the peripheral compartment, respectively; V_1, V_2 are the volumes of the central and peripheral compartments, respectively; F is the fraction of the oral dose D_0 ultimately absorbed.

The concentration C_1 in the central compartment at time t after a single dose is given by the expression

$$C_1 = \frac{k_a F D_0}{V_1}\left(\frac{k_{21} - \alpha}{(k_a - \alpha)(\beta - \alpha)}\right)e^{-\alpha t} + \left(\frac{k_{21} - \beta}{(k_a - \beta)(\alpha - \beta)}\right)e^{-\beta t} + \left(\frac{k_{21} - k_a}{(\alpha - k_a)(\beta - k_a)}\right)e^{-k_a t} \quad (108)$$

The α and β are hybrid rate constants and are related to the specific rate constants of the two-compartment open model by the equations

$$\alpha + \beta = k_{12} + k_{21} + k_e \quad (109)$$

$$\alpha\beta = k_{21}k_e \quad (110)$$

The concentration C_1 in the central compartment after administration of N equal doses at dosing interval d_i is

$$C_1 = \frac{k_a F D_0}{V_1}\left[\left(\frac{k_{21} - \alpha}{(k_a - \alpha)(\beta - \alpha)}\right)e^{-\alpha t'}\left(\frac{1 - e^{-\alpha d_i N}}{1 - e^{-\alpha d_i}}\right) + \left(\frac{k_{21} - \beta}{(k_a - \beta)(\alpha - \beta)}\right)e^{-\beta t'}\left(\frac{1 - e^{-\beta d_i N}}{1 - e^{-\beta d_i}}\right) + \left(\frac{k_{21} - k_a}{(\alpha - k_a)(\beta - k_a)}\right)e^{-k_a t'}\left(\frac{1 - e^{-k_a d_i N}}{1 - e^{-k_a d_i}}\right)\right] \quad (111)$$

In Eq. (111), t' is the time elapsed after administration of the nth dose; that is, $t' = t - (N - 1)d_i$ [136].

On the other hand, the amount of drug in the peripheral compartment A_2 and the cumulative elimination of drug A_e are obtained from Eqs. (112) and (113), respectively.

$$A_2 = k_{12}k_aFD_0\left[\frac{e^{-\alpha t}}{(k_a - \alpha)(\beta - \alpha)} + \frac{e^{-\beta t}}{(k_a - \beta)(\alpha - \beta)} + \frac{e^{-k_a t}}{(\alpha - k_a)(\beta - k_a)}\right] \tag{112}$$

$$A_e = k_a k_e A_e\left[\frac{k_{21} - \alpha}{\alpha(k_a - \alpha)(\beta - \alpha)}\left(i - e^{-\alpha t}\right) + \frac{k_{21} - \beta}{\beta(k_a - \beta)(\alpha - \beta)}\left(1 - e^{-\beta t}\right) + \frac{k_{21} - k_a}{k_a(\alpha - k_a)(\beta - k_a)}\left(1 - e^{-k_a t}\right)\right] \tag{113}$$

where

$$\alpha(+), \beta(-) = \frac{1}{2}\left[(k_{12} + k_{21} + k_e) \pm \sqrt{(k_{12} + k_{21} + k_e)^2 - 4k_{21}k_e}\right] \qquad [114]$$

3. Three-Compartment Open Model

In studying the pharmacokinetics of digoxin, Kramer et al. [137] compared the fit of a two- and three-compartment pharmacokinetic model to experimentally determined serum digoxin concentration-time data following rapid intravenous injections of 1.0 mg of the drug. The three-compartment model is depicted below:

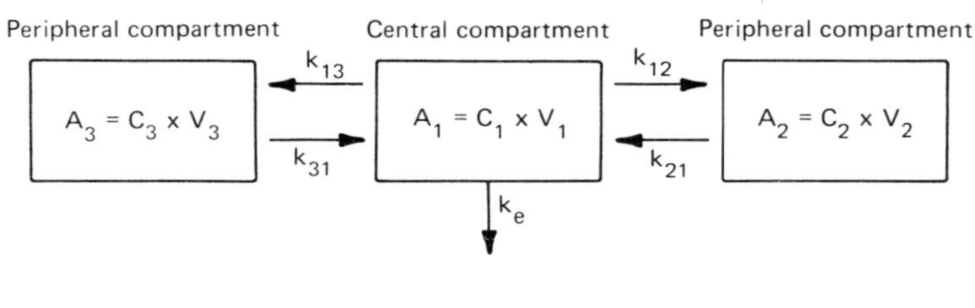

Equation (115) describes the drug concentration in the central compartment after an intravenous bolus injection and was derived for this model by the method used by Benet [138].

$$C_1 = \frac{\text{dose}}{V_1}\left[\left(\frac{(k_{21} - \alpha)(k_{31} - \alpha)}{(\beta - \alpha)(\gamma - \alpha)}\right)e^{-\alpha t} + \left(\frac{(k_{21} - \beta)(k_{31} - \beta)}{(\alpha - \beta)(\gamma - \beta)}\right)e^{-\beta t} + \left(\frac{(k_{21} - \gamma)(k_{31} - \gamma)}{(\alpha - \gamma)(\beta - \gamma)}\right)e^{-\gamma t}\right] \quad (115)$$

For the three-compartment model, the three hybrid constants α, β, and γ may be expressed in terms of the five microconstants. This is accomplished by Kramer et al. in the following manner.

Using Benet's method,

$$(S + \alpha)(S + \beta)(S + \gamma) = (S + E_1)(S + E_2)(S + E_3) - k_{12}k_{21}(S + E_3) - k_{13}k_{31}(S + E_2) \quad (116)$$

where E_i is the sum of the exit rate constants from the ith compartment (i.e., $E_1 = k_{12} + k_{13} + k_e$) and S is a variable introduced with the Laplace transforms used in the derivation. After expanding both sides of Eq. (116) and collecting terms, one obtains the following:

$$\begin{aligned} &S^3 + (\alpha + \beta + \gamma)S^2 + (\alpha\beta + \alpha\gamma + \beta\gamma)S + \alpha\beta\gamma \\ &= S^3 + (E_1 + E_2 + E_3)S^2 + (E_1E_2 + E_1E_3 + E_2E_3 - k_{12}k_{21} \\ &- k_{13}k_{31})S + (E_1E_2E_3 - k_{12}k_{21}E_3 - k_{13}k_{31}E_2) \end{aligned} \quad (117)$$

Solving the resultant cubic equation via a trigonometric solution [139], the numerical values for α, β, and γ can be obtained. Since either side of Eq. (117) is equivalent to

$$S^3 + pS^2 + qS + r = 0 \quad (118)$$

where p, q, and r represent the coefficients of the powers of S, let a and b be defined as

$$a = \frac{1}{3}(3q - p^2) \quad (119)$$

$$b = \frac{1}{27}(2p^3 - 9pq + 27r) \quad (120)$$

The angle θ is calculated using

$$\theta = \arccos\left[-\frac{b}{2} + \left(\frac{-a^3}{27}\right)^{1/2}\right] \tag{121}$$

Finally, the three hybrid constants are determined (by convention, $\alpha > \beta > \gamma$) as follows:

$$-\alpha = \left[2\left(\frac{-a}{3}\right)^{1/2} \cos\left(\frac{\theta}{3} + 240^0\right)\right] - \frac{p}{3} \tag{122}$$

$$-\beta = \left[2\left(\frac{-a}{3}\right)^{1/2} \cos\left(\frac{\theta}{3} + 120^0\right)\right] - \frac{p}{3} \tag{123}$$

$$-\gamma = \left[2\left(\frac{-a}{3}\right)^{1/2} \cos\left(\frac{\theta}{3}\right)\right] - \frac{p}{3} \tag{124}$$

4. Pharmacokinetic Parameters Involved in the Distribution and Elimination of Drugs and Their Metabolites

Mihailova et al. [140] used analog and digital computing techniques to determine the pharmacokinetic parameters involved in the metabolism and excretion of diethylpropion and its metabolites in humans. This is a very good example of a complex problem that is amenable to analysis via the compartmental approach. As noted by Mihailova et al., diethylpropion (A, Fig. 11) is rapidly and completely absorbed from the GI tract and, in humans, its metabolism occurs primarily via two major routes: (a) keto reduction and (b) N-dealkylation; the two routes combine to yield the complex metabolic pattern shown in Fig. 11.

For computer simulation of excretion after dosing with amino ketone E, a scheme such as that indicated in Fig. 12 was found to be the simplest to accommodate experimental results. Tissue storage, designated by subscript T, was included in the compartmental model to describe the excretion of E but was of little or no value for compound F. In Fig. 12, M denotes an unknown metabolite arising from E.

By analogy, a simplified model was proposed for the excretion of amino ketone C after dosing which could be analyzed for rate and cumulative plots of excretion of C and D using a computer program which generated curves and the experimental points indicated in Fig. 13. The complete model for urinary excretion following oral dosing with diethylpropion is shown in Fig. 14.

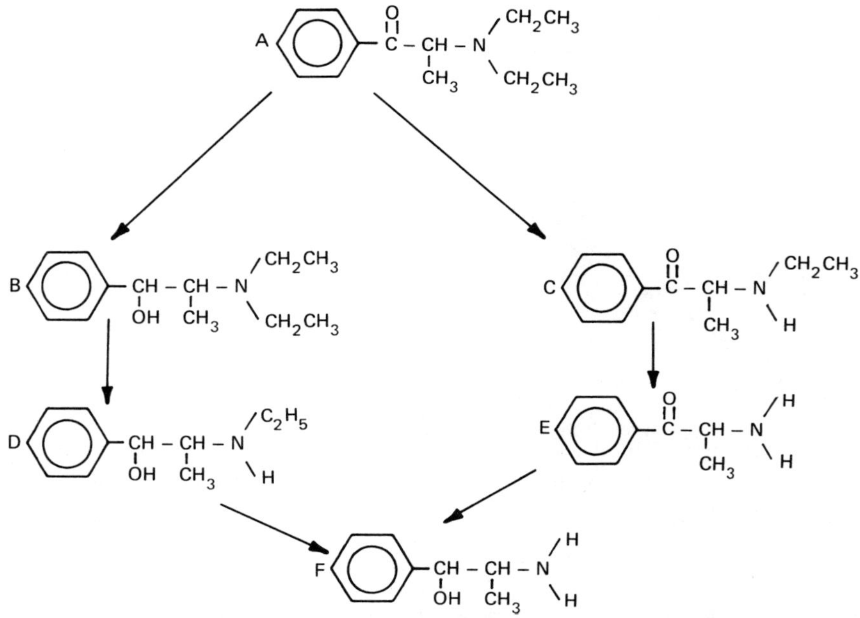

Figure 11 Pathway for the production of the known metabolites of diethylpropion. Adapted from Mihailova et al. [140], courtesy of J. Pharm. Pharmacol.

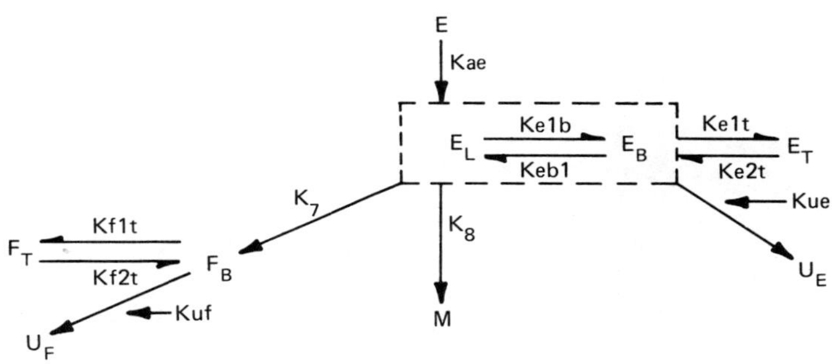

Figure 12 Compartment model for the metabolism and excretion of compounds E and F after dosing with amino ketone E. Adapted from Mihailova et al. [140], courtesy of J. Pharm. Pharmacol.

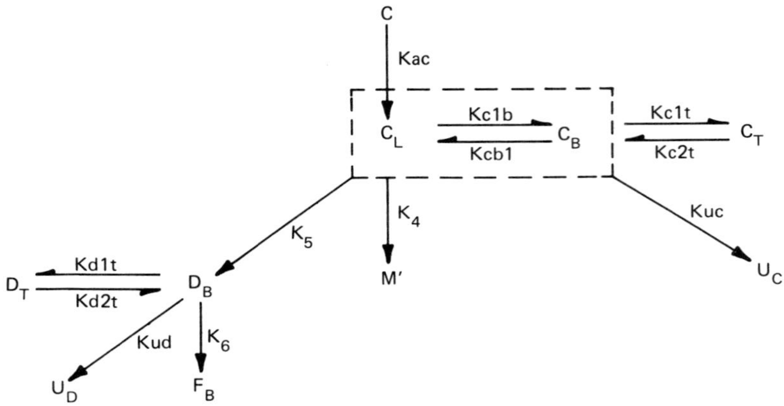

Figure 13 Simplification of compartment model allowing solutions for C and D only. Adapted from Mihailova et al. [140], courtesy of J. Pharm. Pharmacol.

The analog computer program and the definitions of symbols used to solve the models given in Figs. 12-14 are as follows:

1. Equations used for the model shown in Fig. 14.

$$\frac{dA_L}{dt} = KaaA + Kab1A_B - (Ka1b + K_1 + K_2)A_L \tag{125}$$

$$\frac{dA_B}{dt} = Ka1bA_L + Ka2tA_T - (Kab1 + Ka1t + Kua)A_B \tag{126}$$

$$\frac{dA_T}{dt} = Ka1tA_B - Ka2tA_T \tag{127}$$

$$\frac{dU_A}{dt} = KuaA_B \tag{128}$$

$$\frac{dB_B}{dt} = K_1A_L + Kb2tB_T - (Kb1t + Kub + K_3)B_B \tag{129}$$

$$\frac{dB_T}{dt} = Kb1tB_B - Kb2tB_T \tag{130}$$

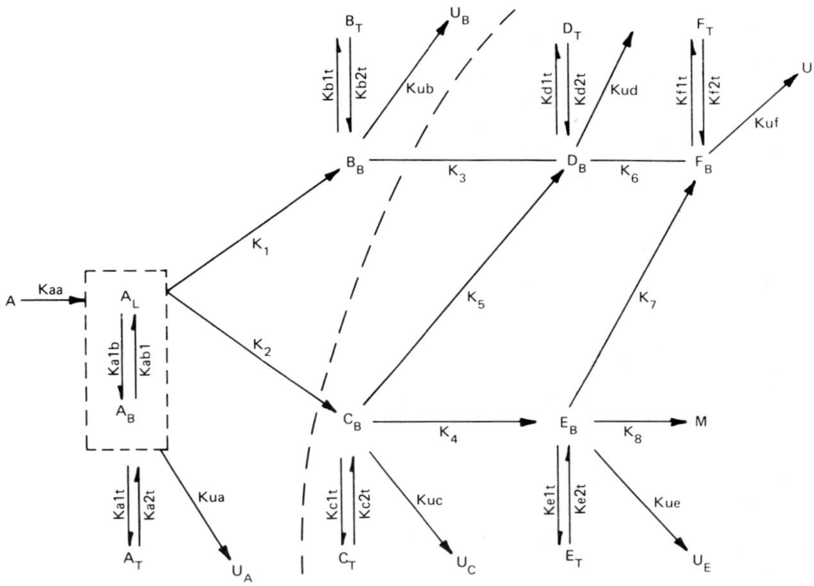

Figure 14 Compartmental model for the metabolism and excretion of diethylpropion and its metabolites A, B, C, D, E, and F following oral dosing with diethylpropion (A). The part of the diagram to the left of the dashed line was used to formulate a simplified model similar to that in Fig. 13 for solution by analog computer. Adapted from Mihailova et al. [140], courtesy of J. Pharm. Pharmacol.

$$\frac{dC_B}{dt} = K_2A_L + Kc2tC_T - (Kc1t + Kuc + K_4 + K_5)C_B \tag{131}$$

$$\frac{dC_T}{dt} = Kc1tC_B - Kc2tC_T \tag{132}$$

$$\frac{dU_B}{dt} = KubB_B \tag{133}$$

$$\frac{dU_C}{dt} = KucC_B \qquad (134)$$

$$\frac{dD_B}{dt} = K_3B_B + K_5C_B = Kd2tD_T - (Kd1t + Kud + K_6)D_B \qquad (135)$$

$$\frac{dD_T}{dt} = Kd1tD_B - Kd2tD_T \qquad (136)$$

$$\frac{dU_D}{dt} = KudD_B \qquad (137)$$

$$\frac{dE_B}{dt} = K_4C_B + Ke2tE_T - (Ke1t + Kue + K_7 + K_8)E_B \qquad (138)$$

$$\frac{dE_T}{dt} = Ke1tE_B - Ke2tE_T \qquad (139)$$

$$\frac{dU_E}{dt} = KueE_B \qquad (140)$$

$$\frac{dF_B}{dt} = K_6D_B + K_7E_B + Kf2tF_T - (Kf1t + Kuf)F_B \qquad (141)$$

$$\frac{dF_T}{dt} = Kf1tF_B - Kf2tF_T \qquad (142)$$

$$\frac{dU_F}{dt} = KufF_B \qquad (143)$$

2. The model shown in Fig. 12 was solved using Eqs. (139), (140), (142), and (143), as well as

$$\frac{dE_L}{dt} = KaeE_L + Keb1E_L + Ke1bE_B - (Ke1b + K_7 + K_8)E_L \qquad (144)$$

$$\frac{dE_B}{dt} = Ke1bE_L + Ke2tE_T - (Keb1 + Ke1t + Kue)E_B \qquad (145)$$

$$\frac{dF_B}{dt} = K_7E_L + Kf2tF_T - (Kf1t + Kuf)F_B \qquad (146)$$

3. The model in Fig. 13 was solved using Eqs. (132), (134), (136), and (137) and

$$\frac{dC_L}{dt} = KacC_L + Ka1bC_B - (Kcb1 + K_4 + K_5)C_L \tag{147}$$

$$\frac{dC_B}{dt} = Kc1bC_L + Kc2tC_T - (Kcb1 + Kc1t + Kuc)C_B \tag{148}$$

$$\frac{dD_B}{dt} = K_5C_L + Kd2tD_t - (Kd1t + Kud + K_6)D_B \tag{149}$$

where

Ka1b, Kc1b, and Ke1b are the respective rate constants for passage of material from "first-pass" compartments A_L, C_L, E_L to "body" compartments A_B, C_B, E_B

Kab1, Kcb1, and Keb1 are the rate constants for the reverse pathways from "first-pass" to "body" compartments

Ka1t, Kb1t, Kc1t, Kd1t, Ke1t, and Kf1t are the respective rate constants for passage of material from "body" compartments A_B, B_B, C_B, D_B, E_B, F_B to "tissue" compartments A_T, B_T, C_T, D_T, E_T, F_T

Ka2t, Kb2t, Kc2t, Kd2t, Ke2t, and Kf2t are the rate constants for the pathways from "tissue" to "body" compartments

K_1, K_2, K_3, K_4, K_5, K_6, K_7, and K_8 are the rate constants for metabolism

Kua, Kub, Kuc, Kud, Kue, and Kuf are the rate constants for urinary excretion

B. Zero-Order Kinetics

In general, rate processes are first-order, but there are exceptions which are noted in cases where a capacity-limited elimination or transport system becomes saturated and behaves as a zero-order rate process [9].

As examples, Goldstein et al. [1] cite (a) the renal tubular secretion of drugs and (b) drug secretion into the bile. As in the case of drug excretion into the bile, a process characterized by a T_m (maximum tubular transport capacity) for each drug, the same principle applies to the renal tubular secretion of drugs. In this instance, if the plasma level of the drug is of such magnitude that

T_m is exeeded, zero-order kinetics will prevail until the concentration falls below the saturation. Below the saturation level, the elimination rate will decrease with decreasing concentration and, at a low enough concentration, the elimination will become always first-order (i.e., proportional to concentration).

In his discussion of zero-order conditions in active transport systems, Notari [9] notes that the type of kinetics observed will be dependent on the conditions. Since the number of transfer sites is finite, and therefore the capacity is limited, it is convenient to consider rate processes under conditions of saturation and nonsaturation. The saturated system, somewhat simpler to envision, can be depicted as a reservoir of drug, far in excess of the capacity of the enzyme system, just sitting and waiting to be transferred. The enzyme system then operates at its maximum capacity and the amount of drug transferred is insignificant as compared to the total transferable drug. Thus the transfer of drug continues at a rate that is equal to the maximum enzyme capacity. This maximum rate would continue as long as the system remained saturated. Since this maximum rate has some given value, the rate is thus a constant. When such a situation exists where the rate is found to be a constant, the process is said to be zero-order. In contrast to a first-order process whose rate decreases with time (being at a maximum at time zero and approaching a zero rate asymptotically), this is not the case for a zero-order process where the rate of transfer will be the same throughout the entire process independent of the time selected for its determination. The difference between a zero-order and first-order process is very evident if concentration versus time is plotted. In the zero-order process, the decrease in concentration per unit of time is the same at any selected time. By graphical means, therefore, one can determine the value of a zero-order rate constant from the linear relationship noted between concentration versus time, the slope of the line being equivalent to the rate constant.

In their discussion of the plateau principle and its application to constant infusions (zero-order absorption, first-order elimination), Goldstein et al. [1] show that the concentration of drug in the system will eventually reach a level at which the output rate is equal to the input rate (i.e., a steady-state system). At this point, this input-output process has a zero-order rate constant. The magnitude of the plateau concentration, however, depends on the infusion rate, the elimination-rate constant for the drug under investigation, and the volume of distribution. If k_e and V_d are fixed, the plateau concentration is related directly to the infusion rate. The time required to reach the plateau concentration is dependent solely on the drug's elimination rate constant k_e and is completely independent of the rate of drug infusion.

If a capacity-limited elimination route is saturated by a large dose, its biological half-life would be dose dependent for a zero-order process. Hence one method to test for saturation of a capacity-limited elimination process is to examine the effect of dose size on the apparent half-life [9].

Notari [9] cites another possibility, that of elimination by two simultaneous routes where one is capacity-limited. Such a situation would exist where elimination of a drug proceeds via urinary excretion and metabolism but, in this instance, the capacity of the enzyme system involved in the metabolic process has been exceeded. Notari represents this situation as

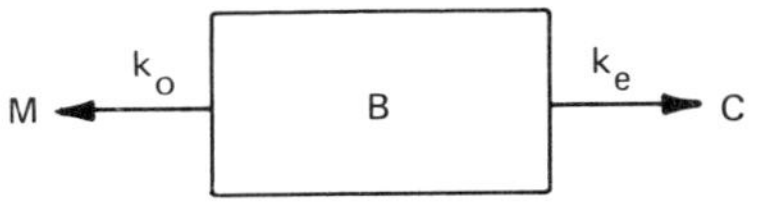

where k_0 is the zero-order rate constant for metabolism and k_e is the first-order rate constant for excretion. As the substrate concentration decreases, a zero-order rate process will predominate over a first-order rate process of equal initial rate.

From the situation diagrammed above, one would note that:

1. The ratio of metabolite to intact drug would decrease with time.
2. The ratio of metabolite to intact drug at a given time would decrease with increasing dose.
3. The percentage metabolized after complete elimination would decrease with increasing dose.
4. The half-life would be independent of dose.

Again, the advantages of intravenous infusion in determining clearance values and obtaining data for capacity-limiting processes is quite obvious. When the rate of infusion becomes equal to the rate of elimination, the plasma concentration C will remain constant. Consequently, there are three distinct advantages in determining clearance values by this approach [9]. The concentration of drug in the blood will remain constant over the determination. The determination of the rate of elimination (amount eliminated per unit time) will no longer involve the assessment of the slope of a curved line since the rate of elimination is equal to k_eC_p, where C_p is a constant. Thus the rate of appearance of drug in the urine will be a constant (i.e., an apparent zero-order process). Since the urinary drug output will be a linear function, the slope of the line can readily be defined in a manner that is relatively accurate compared to a first-order elimination process. An additional advantage lies in the fact that a minor clearance route may be detected by comparing urinary drug output with input rates during steady-state infusion, whereas a minor elimination route might be overlooked in a single-dose study.

Based on a linear single-compartment pharmacokinetic consideration of endogeneous creatinine, Chiou and Hsu [141] derived a new equation to estimate the true renal clearance such that one can determine clinically the creatinine renal clearance of a patient simply by measuring the creatinine serum levels at two different times without the necessity of urine collection. This approach is especially valuable for monitoring the rapidly changing (improving or deteriorating) renal function in all types of patients. Chiou and Hsu used the following single compartment open model [1,142] to describe the disposition kinetics of creatinine in humans:

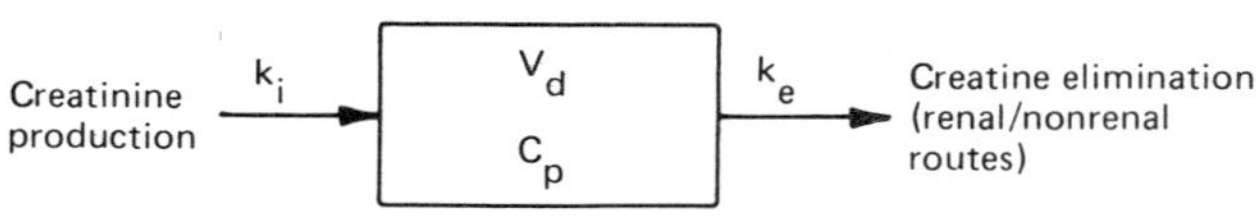

where

k_i = zero-order production rate constant of endogenous creatinine
V_d = apparent volume of distribution of creatinine which is assumed to be constant for a given patient
C_p = serum creatinine concentration
k_e = first-order elimination rate constant from the body which is the summation of k_r, the first-order renal elimination rate constant, and k_{nr}, first-order nonrenal elimination rate constant

To show a change of creatinine concentration with time, the differential equation given below can be written:

$$\frac{dC_p}{dt} = \frac{k_i}{V_d} - k_eC_p \tag{150}$$

Assuming that C_p equals C_{p1} at t = 0 when the first blood sample is drawn for creatinine analysis and k_e (or the renal function) remains constant or unchanged during the time interval Δt before the next sampling time t, it follows that

$$C_p = \frac{k_i - (k_i - k_e V_d C_{p1}) e^{-k_e t}}{k_e V_d} \tag{151}$$

If Eq. (151) is integrated to obtain the area under the curve A_{uc} from time zero to time t with $C_p = C_{p2}$, the total body clearance ($k_e V_d$) can be estimated by

$$k_e V_d = \frac{k_i t}{A_{uc}} + \frac{(C_{p1} - C_{p2}) V_d}{A_{uc}} \tag{152}$$

If A_{uc} is determined by the trapezoidal rule method [143,144], Eq. (152) may be expressed as

$$k_e V_d = \frac{2k_i}{C_{p1} + C_{p2}} + \frac{2V_d (C_{p1} - C_{p2})}{(C_{p1} + C_{p2}) \Delta t} \tag{153}$$

The creatinine renal clearance C_r is determined by subtracting the nonrenal clearance value from the total body clearance such that

$$C_r = k_e V_d - k_{nr} V_d \tag{154}$$

or, by substitution,

$$C_r = \frac{2k_i}{C_{p1} + C_{p2}} + \frac{2V_d (C_{p1} - C_{p2})}{(C_{p1} + C_{p2}) \Delta t} - K_{nr} V_d \tag{155}$$

According to Chiou and Hsu, if it is assumed that the values of k_i, V_d, and k_{nr} are known in a patient (these values also unaffected by the renal failure [142]), it is possible to predict C_r and hence the degree of renal failure or improvement using Eq. (155) by simply measuring serum creatinine levels at two different times. In the absence of individually measured values for

k_i, V_d, and k_{nr} for each patient, average data obtained from an appropriate population could be used for the clinical application.

In their model, the urinary excretion rate of endogeneous creatine was assumed to be approximately equal to its production rate in the body. The values of V_d were either measured or assumed to be nearly equal to the space of the total body water. It is believed that a value equivalent to nearly 60% of body weight could be used as a clinical approximation for most patients with normal body size.

Furthermore, based on Eq. (151), the stable serum creatine concentration $C_{p(s)}$ of patients with chronic renal failure can be given by

$$C_{p(s)} = \frac{k_i}{k_e V_d} = \frac{k_i}{k_r V_d + k_{nr} V_d} \tag{156}$$

which predicts a linear relationship between $1/C_{p(s)}$ and renal creatinine clearance $k_r V_d$; such a relationship having been observed in studies reported by other investigators [145,146].

C. Biologic or Elimination Half-Life

On the basis of drug concentrations in blood, the biological half-life $t_{1/2}$ may be defined as the time required for the body to eliminate one-half of the drug which it contains. For first-order elimination processes, Eq. (106) can be equated to 0.5 (half of the original dose of drug administered), so that

$$t_{1/2} = \frac{0.693}{k_e} \tag{157}$$

If the elimination kinetics behaves as a first-order reaction, one would obtain an exponential curve as indicated in Fig. 15 where the rate of elimination of drug from the body is proportional to its concentration at any given time. Conceptually, if expressed in other terms, this means that the change per unit time is a function of the drug concentration. If the disappearance curve is plotted on a semilog scale as shown in Fig. 16 for two different drugs (A and B), a straight line will result. From these plots, k_e values (elimination rate constants) for each drug may be determined; k_e being equivalent to the slope of each straight line. In the example given, $t_{1/2}$ for drug B is twice

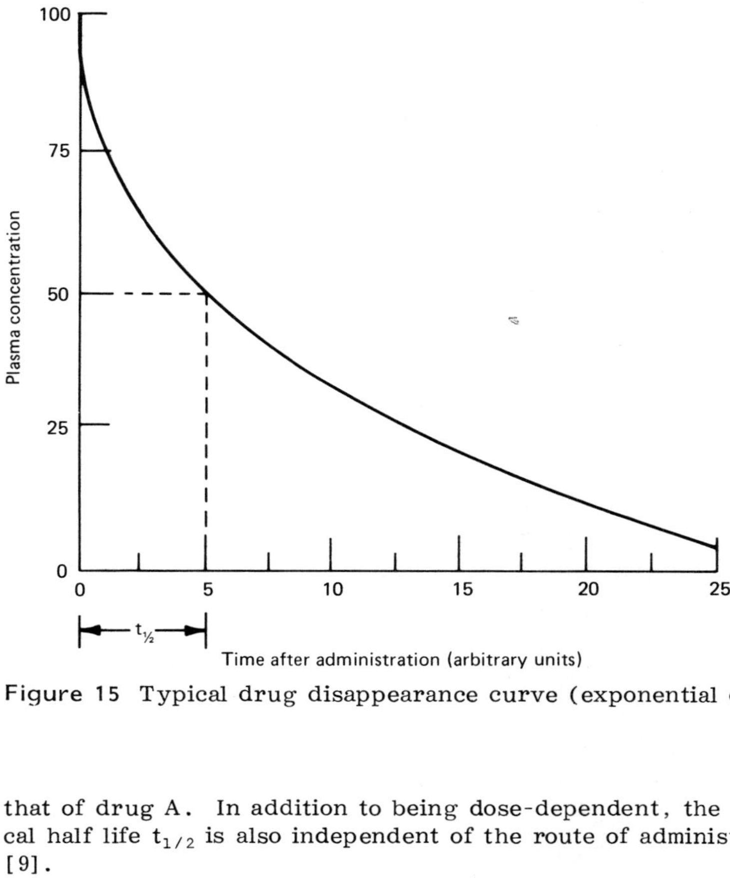

Figure 15 Typical drug disappearance curve (exponential decay).

that of drug A. In addition to being dose-dependent, the biological half life $t_{1/2}$ is also independent of the route of administration [9].

As pointed out by Goth [6], erroneous ideas in therapeutics may result from a failure to appreciate the exponential nature of drug disappearance. For example, if a drug has a short duration of action, doubling the dose will not double the duration of its effect. To prolong the biological action of a drug, several methods are available: (a) frequent administration, (b) slowing down or decreasing its absorption rate, (c) interfering with its renal excretion, or (d) inhibiting drug metabolism.

It should be noted that a relationship exists between elimination half-life and total body clearance. By definition, total body clearance (k_eV_d) is realistically the sum of the drug's renal and metabolic clearance:

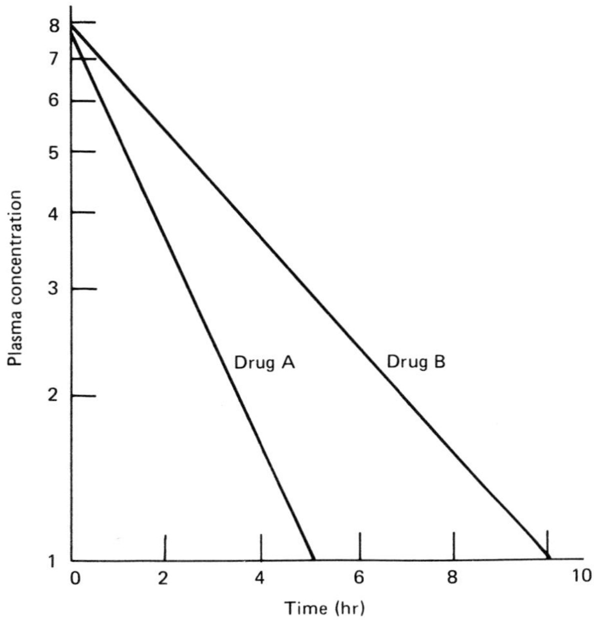

Figure 16 Semilog plots of drug disappearance curves.

$$k_e V_d = k_r V_d \text{ (renal clearance)}$$

$$+ k_m V_d \text{ (metabolic clearance)} \qquad (158)$$

where k_r and k_m are the first-order renal excretion and metabolic rate constants, respectively. For a given clearance, the larger the V_d, the smaller the k_e; hence the $t_{1/2}$ increases [45]. Thus, following its administration, the $t_{1/2}$ of a drug depends on its clearance and the manner in which it becomes distributed within the body.

As a practical example, let us consider the gas chromatographic method developed by Graham and Bolt [147] to determine the levels of diphenylpyraline in urine from which the biological half-life was subsequently determined from the excretion of the unchanged drug. In their study, the $t_{1/2}$ of diphenylpyraline was calculated by two methods: (a) from the cumulative urinary excretion of

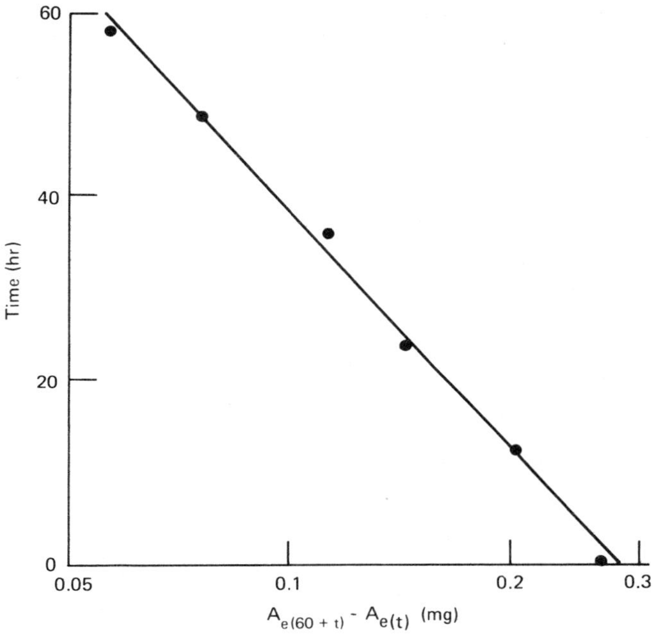

Figure 17 Guggenheim plot. Difference between cumulative excretion at 60 + t and t hours. $A_e(60 + t) - A_{e(t)}$ plotted versus time. Point at zero time excluded from regression analysis. Adapted from Graham and Bolt [147], courtesy of J. of Pharmacokinetics and Biopharmaceutics.

unchanged drug by the method of Guggenheim [148], and (b) from semilog plots of the rate of excretion versus time. Both procedures yielded similar values. However, using the Guggenheim method, one must solve a regression analysis of the relationship between $\ln(A_{e(60+t)} - A_{e(t)})$ and t, where $A_{e(60+t)}$ and $A_{e(t)}$ are the cumulative amounts of unchanged drug excreted at 60 + t and t hours after dosage. Applicable only to monoexponential functions, the biological half-life was estimated from t values ranging from 12 to 60 hr. A typical Guggenheim plot is shown in Fig. 17, where one notes that the agreement between the experimental points and the regression line is good, consistent with first-order elimination of diphenylpyraline.

To determine the amount of drug excreted unchanged, $A_{e(u)}$, Graham and Bolt used a method analogous to that reported by Beckett and Rowland [149]; $A_{e(u)}$ is given by

$$A_{e(u)} = A_{e(12)} + \frac{A_{e(120-12)}}{1 - e^{-108k_e}} \tag{159}$$

where

$A_{e(12)}$ = amount of diphenylpyraline excreted up to 12 hr after dosage
$A_{e(120-12)}$ = amount excreted between 12 and 120 hr after dosage
k_e = apparent rate constant of elimination $(0.693/t_{1/2})$

Their calculated $A_{e(u)}$ values clearly indicated that only small fractions of the oral doses of diphenylpyraline were excreted unchanged.

5

Mechanisms of Drug Action

I. DRUG RECEPTORS

For a specific drug, its site of action is that part of the body—organ, tissue, or cell—where drug-site interactions initiate the chain of events that culminate in an observed effect. In the body, the site of action may be either extracellular, intracellular, or at the cellular surface. The cell compartment or functionally important molecule in the living organism directly involved with physicochemical drug interactions is usually termed a receptor. Hence a receptor may be defined as that functional or organized group of atoms of a cell component with which a drug combines reversibly to initiate a response [70]. In contrast to drug-receptor binding that initiates an action or a response, there are also those interactions that are only binding in nature such as the binding of drugs to plasma proteins, cell proteins, and enzymes concerned primarily with the transport and biotransformation of drugs. These have been characterized and designated in many ways by such terms as secondary receptors, silent receptors, storage sites, drug receptors, or sites of loss.

II. BINDING FORCES OF DRUG-RECEPTOR INTERACTIONS

In general, the bonds formed between a receptor and a drug are rather weak such that the effects produced are reversible; these weak bonds for the most part being (a) ionic, (b) polar, (c) hydrogen, (d) hydrophobic, or (e) van der Waals in nature. Although seldom formed and difficult to cleave, it is possible for a drug and receptor to establish a covalent bond by the sharing of a pair of electrons between the two atoms involved, atoms such as carbon, hydrogen, nitrogen, and oxygen. The weaker bonds are very similar to those that stabilize protein structures as depicted in Fig. 18. Korolkovas [116] notes that bond strength depends on the distance separating the "interactive" atoms and the amount of free energy released in bond formation. The greater the free-energy change, the greater the proportion of atoms in the bonded form. This is readily seen from the expression

$$K_{eq} = e^{-\Delta G°/RT} \tag{160}$$

or

$$\Delta G° = -RT \ln K_{eq} \tag{161}$$

where

K_{eq} = equilibrium constant
$\Delta G°$ = standard free energy
R = gas constant
T = absolute temperature

Albert [151] has compiled average interaction energies (-kcal/mole) for various bond types; these being (a) covalent, 40-110; (b) reinforced ionic bond, 10; (c) ionic bond, 5; (d) ion-dipole bond, 1-7; (e) dipole-dipole, 1-7; (f) hydrogen bond, 1-7; (g) charge transfer, 1-7; (h) hydrophobic interaction, 1; and (i) van der Waals forces, 0.5-1.0.

Unlike most drug-receptor interactions, the covalent binding of receptors and drugs results in stable, long-lasting complexes. Such drugs participating in covalent bond formation include alkylating agents which form reactive intermediate carbonium ions by virtue of the strong electron-attracting power of the group to which the alkyl residue is initially attached [152-154]; organophosphorus insecticides, phosphates, and carbamates [155-157] which combine with cholinesterase enzyme sites; arsenicals, mercurials, and antimonials that interact with two sulfhydryl groups or with two different receptor sites, one being a -SH group and the other an imidazole ring, a phenolic hydroxy group, an amino group, or a carboxy group [158]. The coordinate covalent bond is formed when both electrons of the electron pair that form the

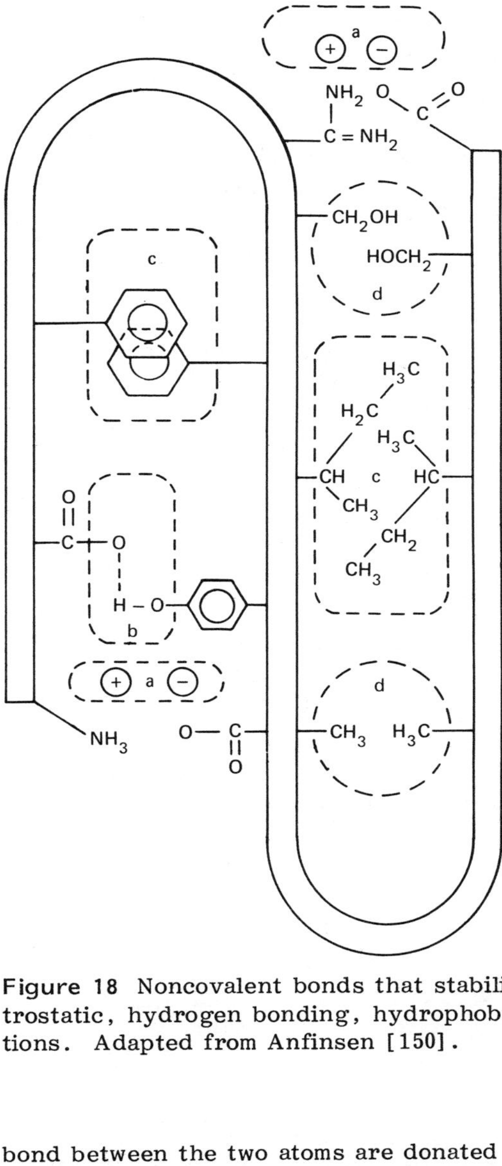

Figure 18 Noncovalent bonds that stabilize proteins: ionic or electrostatic, hydrogen bonding, hydrophobic, van der Waals interactions. Adapted from Anfinsen [150].

bond between the two atoms are donated by the same atom. In biological systems, the donor atom is usually sulfur, nitrogen, or oxygen. The atoms that share donated electron pairs in coordination complexes are usually metallic cations (Na^+, Mg^+, K^+, Ca^{2+}, Cu^+,

Cu^{2+}, and Zn^{2+} as well as transition metals Mn, Fe, and Co). In the covalent bond category of reactions must also be included chelate complexes whose stability depends on the nature of the chelating agent and the complexed atom.

As for compounds exhibiting electrostatic or ionic attraction, these result from coulombic forces between oppositely charged ions, q_1 and q_2, separated by a distance r. The energy E of the ionic bond in a medium with a dielectric constant D may be expressed as

$$E = \frac{q_1 q_2}{Dr} \tag{162}$$

In many drugs, cationic and anionic groups are present which are capable of forming ionic bonds with oppositely charged receptors such as the anionic carboxyls of proteins; aspartic and glutamic acids; the sulfhydryl groups of cysteine; the anionic phosphoryl residues of nucleic acids or the cationic groups in lysine, arginine, histidine, and amide $-NH_2$ of glutamine, and asparagine and terminal amino $-NH_2$ of proteins [1]. On the other hand, the ionic bond is said to be "reinforced" when its bond strength is enhanced by the simultaneous presence of other interactions (hydrophobic, charge transfer, hydrogen bonding, dipole, van der Waals) [151].

A hydrogen bridge or bond is a special ionic bond that arises from a proton's ability to accept an electron pair in part from each of two electron donor atoms such as nitrogen or oxygen. In addition to being highly directional, the strength of hydrogen bonding is inversely proportional to the distance between the atoms [159]. In general, bond strength increases with increasing electronegativity of the atoms that are bridged by the hydrogen.

In biological systems, the most important hydrogen bonds encountered are those formed between hydrogen atoms covalently bonded to oxygen (O-H) or nitrogen (N-H) and drugs containing such groups as -OH, carbonyl, amino, and imino, which may, however, act either as donors or as acceptors of protons. Furthermore, in addition to drastically altering drug solubility, intramolecular bonding can substantially influence other physicochemical properties, which, in turn, have an effect on the drug's pharmodynamic activity.

The charge transfer bond is a special case of donor-receptor complex formation via hydrogen bonding resulting from an electrostatic attraction between electron-donating and electron-accepting molecules. Regarded as ionic pairs, Baker [160] has classified both donors and acceptors into two categories for a particular charge-transfer complex, the donors being either rich in π electrons or having an unshared electron pair, whereas acceptors are those containing strongly electron-withdrawing groups or weakly acidic hydrogens. The electron-donating ability of a given compound can be deduced spectrophotometrically from the absorption wavelength attributed to the charge-transfer complex, an increase in wavelength indicating enhanced electron-donating ability.

Differences in electronegativities between carbon and other atoms result in an asymmetric distribution of electrons (i.e., the formation of dipoles as is commonly noted in ester, carbonyl, ether, amide, nitrile, and other groups that carry partial positive and negative charges) [161]. Korolkovas [116] shows that the energy of interaction (E_i) between one fixed charge and a dipole is given by [162].

$$E_i = \frac{Ne\mu \cos \theta}{D(r^2 - d^2)} \tag{163}$$

where

N = Avogadro's number
e = electrical charge
μ = dipole moment
θ = angle between the line that binds the fixed charge to the dipole center and the dipole direction
D = effective dielectric constant of the separating medium
r = distance between the fixed charge and the dipole center
d = length of the dipolar bond

For two polar molecules with dipole moments μ, separated by a distance r, the dipole-dipole energy of attraction $E_{d\text{-}d}$, for values of r large compared with the charge separation in the dipole, can be determined by the expression [163]

$$E_{d\text{-}d} = -\frac{2}{3}\frac{\mu^4}{r^6}\frac{1}{kT} \tag{164}$$

At higher temperatures, the thermal motion of the molecule competes with the tendency toward favorable orientations; thus the dipole-dipole attraction becomes less important. On the other hand, the attractive energy E_{ind} resulting from the dipole-induced dipole, or induction, effect for a pair of molecules with dipole moments μ and polarizabilities α is

$$E_{ind} = -\frac{2\alpha\mu^2}{r^6} \tag{165}$$

where α is defined here as the dipole moment induced in a molecule by an electric field of 1 V/cm.

The van der Waals bond is a very weak interaction between dipoles or induced dipoles, often between similar atoms. Although short in duration, the intensity of van der Waals forces E_V may be derived from

$$E_V = \frac{-A}{r^6} + \frac{B}{r^{12}} \tag{166}$$

where

r = distance that separates the centers of the two atoms
A, B = constants characteristic of the atom's electronic structures [164]

The force of attraction is inversely proportional to the seventh power of distance and, for this reason, a slight increase in interatomic distance decreases these forces rapidly. Mantione [165] states that four specific factors must be considered in the determination of van der Waals forces (known also as London dispersion forces): (a) E_{pol}, polarization energy; (b) E_{disp}, dispersion energy; (c) E_{el}, electrostatic interaction between net atomic charges; and (d) E_{rep}, a repulsion term. The greater the polarizability of the groups that approach one another, the greater their interaction because the maximal forces between the different groups are proportional to the product of their polarizabilities [116].

Finally, the weak hydrophobic interactions between two nonpolar chain surfaces resulting from a disorientation or displacement of water molecules previously occupying sites adjacent to the surfaces are the consequence of a concurrent decrease and increase in the system's free energy and entropy, respectively. This stabilization or association between the two nonpolar groups (one an integral part of a macromolecule and the other incorporated into the therapeutic agent's structure) is called a hydrophobic bond or interaction, a condition that is highly favorable to complex formation.

III. DRUG STRUCTURE-BIOLOGICAL ACTIVITY RELATIONSHIP

The interaction of a drug and its receptor (the consequence of a reversible physicochemical reaction) provides the stimulus or response which results in the effect observed, either as a biochemical or physiologic change or as the disappearance or appearance of clinical symptoms [4].

$$D + R \underset{k_2}{\overset{k_1}{\rightleftharpoons}} DR \xrightarrow{k_3} \text{effect} + D + R \qquad (167)$$

where

D = agonist, a drug with stimulant properties
R = receptor
DR = drug-receptor complex

In Eq. (167), drug response or effect is directly related to the drug-receptor complex concentration and, at equilibrium, its potency (biological activity per unit weight) has been explained by its affinity for the receptor. Whereas the dissociation constant K_D of drug D can be expressed by the law of mass action in a manner similar to that described previously for plasma protein binding of drugs [Eq. (66)], its binding capacity or affinity is the reciprocal of the dissociation constant $1/K_D$, which is the same as its association constant K_A [see Eq. (59)]. On the other hand, drug effectiveness or capacity to stimulate an action after fixation is expressed by the term "efficacy" or "intrinsic activity"; this designated as e', which is the ratio of the response or stimulus produced by the drug S_D to the maximum stimulus or effect obtainable with the drug S_M. Hence intrinsic activity or efficacy is

$$e' = \frac{S_D}{S_M} \qquad (168)$$

or, as defined by Stephenson [166], the stimulus based on the occupancy theory may be expressed as

$$S_D = e'p_D \qquad (169)$$

where p_D is the fraction of the total receptors occupied by D (drug). The theories of drug action will be more fully discussed in a later section of this book. It will be shown that log dose-response curves may shift due to the varying affinities exhibited by drugs for receptors, whereas varying efficacies or intrinsic activities are readily deduced from the observed variations in the maximal height of the curves.

Whereas agonists provide the stimulus for a biological response, agonist-receptor complex formation can be inhibited by other drugs, known as competitive or noncompetitive antagonists. It is classified as being competitive when it can be displaced by an excess of agonist from receptor sites for which it is "competing" with the active drug and with which it combines reversibly (i.e., the receptors). In this context, it can be said that the antagonist has affinity but lacks intrinsic activity (efficacy). From the agonist's log dose-response curve, one will note a parallel displacement to the right in the presence of a competitive antagonist (i.e., a change in the agonist's affinity for the receptor) without a change in either the curve's shape or maximum response.

On the other hand, a noncompetitive antagonist produces different effects on log dose-response curves that are readily recognized: (a) a decrease in its maximum height (i.e., the inability of an excess of agonist to displace the antagonist from receptor sites), (b) a reduction of the slope of the curve in proportion to the magnitude of the noncompetitive antagonism produced, and (c) a concurrent shift to the right of the resulting agonist-antagonist curve [4].

In addition to agonists and antagonists, there are also those drugs classified as partial antagonists that possess affinity and some efficacy. Occupying receptor sites and producing only weak effects on their own, they nevertheless inhibit or prevent access of strong agonists to these receptor sites.

For the drug to get to its site of action as depicted in Fig. 1, it is subjected to a variety of biological processes (administration, dissolution, distribution, absorption, metabolism, excretion, reabsorption, etc.). Hence it must have certain physicochemical properties to pass various biological barriers as well as chemical specificity and structure-action relationships [167].

Depending on the mechanism or mode of drug action, drugs have been essentially classified into two categories: (a) structurally specific or (b) structurally nonspecific [71,168,169].

A. Structurally Specific Drugs

A drug is considered to be structurally specific if its biological response is attributable to its chemical structure (implied is the effect due to such factors as resonance, induction, isoterism, oxidation-reduction potentials, types of chemical bonds, electronic distribution, and so on, as well as spatial conformation). In contrast to structurally nonspecific drugs, their biological activity is not dependent on the degree or relative saturation at some cellular phase, that is, its thermodynamic activity as proposed by Ferguson [170]. From basic thermodynamic principles, the change in free energy for a constant-temperature process is given by

$$\Delta G = \Delta H \ T \ \Delta S \tag{170}$$

where

ΔG = change in Gibbs free energy
ΔH = change in enthalpy or total energy of the system
ΔS = change in entropy
T = absolute temperature

For a reaction

$$aA + bB \rightleftharpoons cC + dD \tag{171}$$

where a, b, c, and d are the number of reactants A and B and products C and D, respectively, a point of equilibrium is reached at which no further net chemical change occurs and where the free energy of the system is a minimum. The equilibrium constant K_{eq} is therefore given by

$$K_{eq} = \frac{[C]^c[D]^d}{[A]^a[B]^b} \tag{172}$$

In turn, the standard free-energy change $\Delta G°$ must be considered when all the reactants and products are in their normal states (i.e., when $\Delta G = \Delta G°$). Being a mathematical function of its equilibrium constant, it can be expressed for our example as

$$\Delta G° = -RT \ln \frac{[C]^c[D]^d}{[A]^a[B]^b} = -RT \ln K_{eq} \tag{173}$$

where R is the gas constant. It follows therefore that the free energy is given by

$$\Delta G = \Delta G° + RT \ln K_{eq} \tag{174}$$

At equilibrium when $\Delta G = 0$, Eq. (174) becomes Eq. (173).

From Eq. (174), Ferguson defined the thermodynamic activity a_t of a drug in a gas or in a solution as

$$\ln a_t = \frac{\bar{G} - \bar{G}_0}{RT} \tag{175}$$

where $\ln a_t$ is the natural logarithm function of the difference in the partial molal free energy in a particular state ($\bar{G}$) and its normal state ($\bar{G}_0$).

In the case of volatile drugs, the thermodynamic activity may be determined from the ratio p_t/p_s, p_t and p_s being the drug's partial pressure in solution and its saturated vapor pressure at the experimental temperature, respectively. For nonvolatile therapeutic agents, the ratio S_t/S_0 is used to determine the thermodynamic activity, S_0 and S_t designating the drug's solubility and molar concentration, respectively.

Having a low thermodynamic activity, structurally specific drugs are usually more effective at lower concentrations than those that are structurally nonspecific. Furthermore, most drugs that belong to the same class have common structural characteristics which orientate functional groups into similar spatial arrangements, complementary to that of the receptor. Based on such specific structural entities, Burger [171] has compiled data showing biological activities associated with certain ring systems (indole and condensed systems, phenothiazine, isoquinoline, pyrazolone, quinolines and condensed systems, pyrimidines and condensed systems, as well as purines and pteridines). Using a congeric series of chemically related drugs, inferences can be made about the precise mechanism of combination of the receptor's surface with the drug by monitoring the biological potency. From such a study, conclusions can be drawn about the nature of the drug-receptor interaction and the conformation of the receptor's surface [116,172-177]. For example, Ariens [178,179] distinguishes between the pharmacological effects of biofunctional (those determining biological activity) and chemofunctional (the bond types or binding forces associated with drug-receptor interactions) moieties. In turn, Ariens [178] categorizes biofunctional moieties into several subdivisions:

1. Critical and noncritical groups which either participate in drug-receptor complex formation (critical) or do not interact at all with the receptor (noncritical).
2. Isoteric and bioisoteric groups: (a) isoterism is a term used by Langmuir [180] to describe compounds or atomic groupings having the same number and arrangement of electrons, whereas (b) bioisoterism as conceived by Friedman [181] is applied to those compounds which "fit the broadest definition for isosteres and have the same type of biological activity."
3. Carrier groups: the introduction of an appropriate carrier group via the principle of drug latentiation [182] to (a) prolong or shorten the duration of drug, (b) permit drug localization in a particular target tissue or cell

through regulation and penetration of the drug within the body, (c) minimize difficulties arising from drug formulation processes (either by modifying its physical properties or combining it with an inert material), or (d) reduce toxic and side effects.
4. Haptophoric and pharmacophoric groups: (a) haptophoric groups assist in drug-receptor binding, whereas (b) pharmacophoric moieties are those responsible for biological action.
5. Vulnerable groups: the introduction of a group to enhance or inhibit a drug's resistance or susceptibility to enzymatic action.

Since a degree of complementarity between drug and receptor is essential for interaction, stereochemistry of drugs plays an important role in whether or not the isomeric compounds will exhibit pharmacological activity. Let us consider, for example, the two optical isomers, d- and l-epinephrine, which have the same functional groups and therefore may undergo similar chemical reactions. Nevertheless, in spite of the similarity, these compounds have different physical properties, different spatial arrangements of their substituent groups, different rates of reactivity, and different biological properties. As postulated by Easson and Stedman [183], d-epinephrine owes its activity to the ease with which it can be attached to the receptor surface as shown in Fig. 19. Here the active isomer makes a three-point contact with its receptor through three groups: (a) the hydroxyl group in the side chain, (b) the amino group, and (c) the aromatic benzene ring with its two phenolic -OH groups.

In addition to the spatial arrangement of groups in a molecule, interatomic distances and electronic distribution configurations are also very important in drug design, as indicated by the examples given in Fig. 20. As noted by Gourley [70], it is obvious that attempts to define possible receptor structures (i.e., receptor topography) are at best speculative. Nevertheless, these are being pursued rather successfully using antimetabolites, covalent labeling, and compounds of rigid structure that "fit well" into the type of receptor sites depicted in Fig. 20. Such approaches have been used to describe in schematic fashion the receptors for narcotic analgesics, acetylcholine, anti-inflammatory agents, adrenergic amines, local anesthetics, histamine, polycyclic carcinogenics, phenothiazine tranquilizers, diuretics, serotonin, monoamine oxidase inhibitors, steroids, and so on [116].

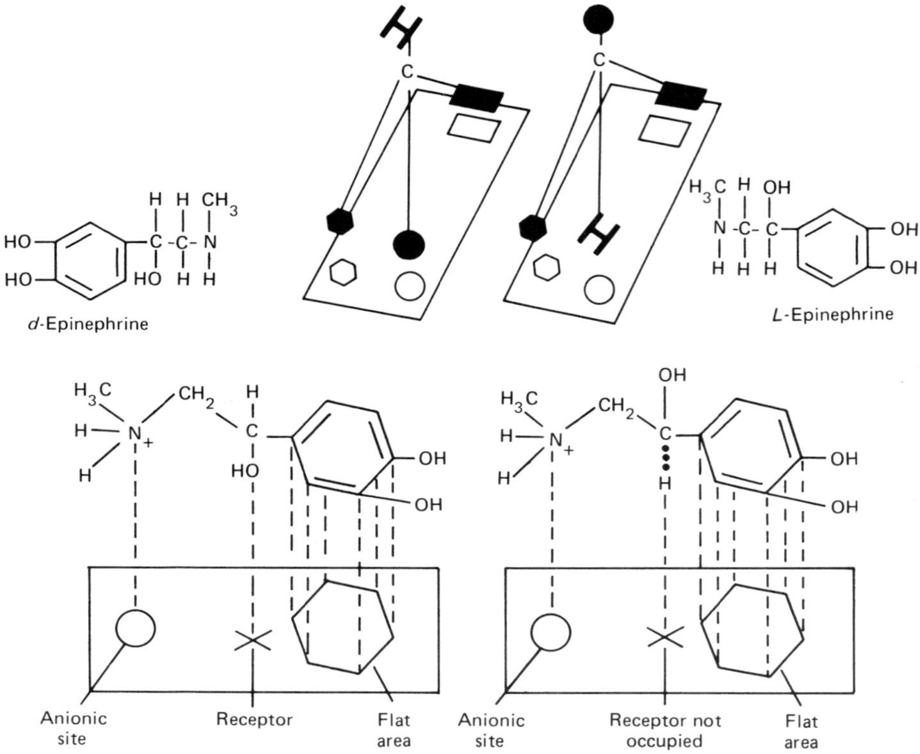

Figure 19 Interaction of isomeric drugs with a complementary receptor surface. Adapted from Albert [151] with Daniels and Jorgensen [71].

B. Structurally Nonspecific Drugs

These are compounds whose pharmacological properties cannot be attributed to their chemical structure but whose potency in part is related to physicochemical properties such as thermodynamic activity, oil-to-gas coefficient, adsorption, solubility, oxidation-reduction potential, pK_a, permeability, complex formation or chelation, membrane depolarization, protein coagulation, van der Waals forces, molecular weight, vapor pressure, dissociation pressure of hydrate crystals, and so on, this category including such

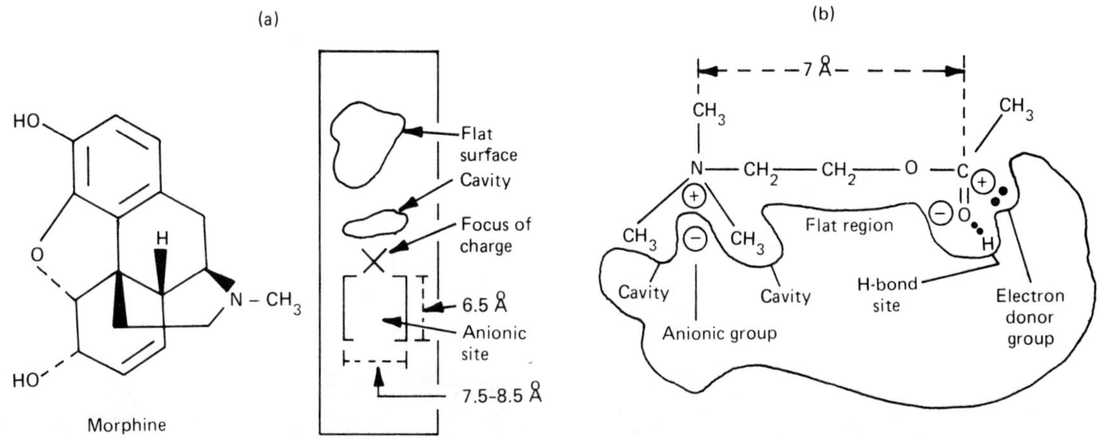

Figure 20 Hypothetical relationships between acetylcholine (b) and morphine (a) and their respective complementary receptors. Adapted from Beckett [169] and Goldstein et al. [1].

therapeutic agents as anesthetics, antacids, adsorbents, astringents, cathartics, chemical antidotes, demulcents, lubricants, osmotic diuretics, plasma expanders, and so on [70].

IV. THEORIES OF DRUG ACTION

To describe drug action, many theories have been proposed of which the four most predominant ones are (a) occupancy, (b) rate, (c) induced fit, and (d) macromolecular perturbation.

A. Occupancy Theory

Originally proposed by Clark [184] and Gaddum [185,186], the theory states that the magnitude of the drug response is proportional to the occupancy of the receptors by drug molecules with maximum response achieved when all receptor sites have been "occupied." Using simple mass law principles such as those described for drug-protein binding, drug-receptor interactions follow Langmuir's adsorption isotherm and may be expressed by Eq. (167), where k_1 and k_2 are rate constants of adsorption and desorption, respectively, k_3 a proportionality factor, and E the pharmacological response or effect. Since drug effect depends on the degree of receptor sites occupied or the concentration of the drug-receptor complex DR, drug response or effect can therefore be given by

$$E = k_3[DR] \tag{176}$$

At equilibrium, the dissociation constant K_D of the drug-receptor complex can be expressed as

$$K_D = \frac{k_2}{k_1} = \frac{[D][R]}{[DR]} \tag{177}$$

and if the total receptor sites R_{tot} is equivalent to

$$[R_{tot}] = [R] + [DR] \tag{178}$$

then, by substitution,

$$K_D = \frac{\left(\left[R_{tot}\right] - [DR]\right)\ [D]}{[DR]} \tag{179}$$

or, by rearrangement,

$$\frac{[DR]}{[R_{tot}]} = \frac{[D]}{[D] + K_D} = \frac{1}{1 + (K_D/[D])} \quad (180)$$

According to the occupancy theory, the maximum response E_{max} will be attained when all receptor sites are occupied by drug molecules. When this requirement is fulfilled, the value of E_{max} can be determined by

$$E_{max} = k_3 [R_{tot}] \quad (181)$$

Dividing Eq. (176) by Eq. (181), we have

$$\frac{E}{E_{max}} = \frac{[DR]}{[R_{tot}]} \quad (182)$$

Substituting for $[DR]/[R_{tot}]$ yields

$$E = \frac{E_{max}[D]}{[D] + K_D} \quad (183)$$

which is identical to the Michaelis-Menten equation, given below:

$$V = \frac{V_{max}[S]}{K_m[S] + K_m} \quad (184)$$

where

V = enzymatic reaction rate
V_{max} = maximum reaction rate
K_m = Michaelis-Menten dissociation constant

If one plots E versus [D] [Eq. (183)] and V versus [S] [Eq. (184)], similar Langmuir-type graphs are obtained, as indicated in Fig. 21.

Since drug-receptor interaction is not the same for all drugs, Ariens [187,188] and Stephenson [166] modified the "occupancy theory" into two phases: (a) the formation of the drug-receptor complex and (b) the production of effect. The concept developed by Ariens-Stephenson retains the assumption that response is

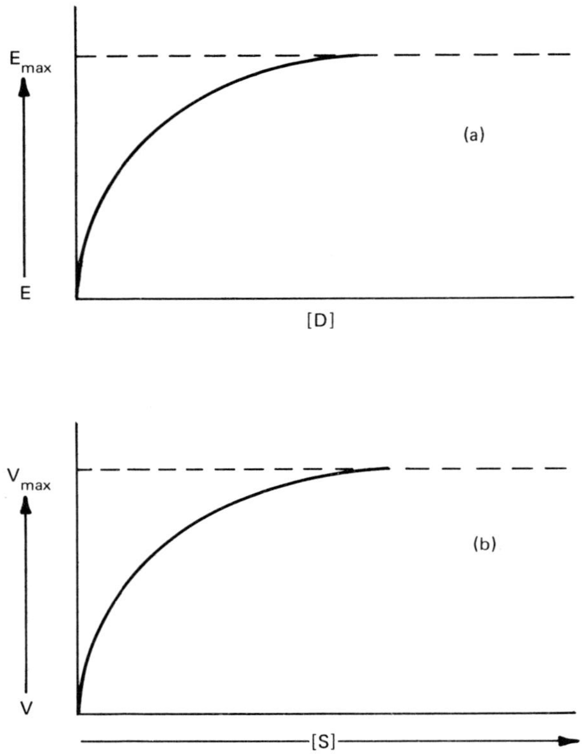

Figure 21 (a) Drug response or effect versus drug concentration; (b) enzymatic reaction rate versus substrate concentration.

related to the concentration of the drug-receptor complex but also postulates that any drug may have structural features which contribute independently to the affinity for the receptor, and to the efficacy with which the drug-receptor combination initiates the response (intrinsic activity or efficacy). The drug's intrinsic activity or efficacy determines the complex's ability to produce the biological response or effect.

Ariens proposed the following expression for the response produced by the drug E_D where α is the intrinsic activity.

$$E_D = [DR] = \frac{[R_{tot}]}{1 + (K_D/[D])} \tag{185}$$

On the other hand, if E is a positive function of the stimulus S_D (this equivalent to the efficacy e multiplied by the fraction of receptor sites that are occupied by the drug), then the Stephenson equation may be used, given by

$$E = f(S_D) = f\left(\frac{e[DR]}{[R_{tot}]}\right) = f\left(\frac{e}{1 + (K_D/[D])}\right) \tag{186}$$

More recently, Ariens et al. [189,190] have proposed an equation quite similar to that derived by Stephenson, this being

$$\frac{E_D}{E_{max}} = f\left(\frac{S_D}{S_{max}}\right) = f\left(\frac{\alpha[DR]}{[R_{tot}]}\right) \tag{187}$$

If one inverts Eq. (183), one obtains the equation for a straight line:

$$\frac{1}{E} = \left(\frac{K_D}{E_{max}}\right)\frac{1}{[D]} + \frac{1}{E_{max}} \tag{188}$$

when 1/E is plotted against 1/[D]. Known as a double-reciprocal plot, it can be used for the analysis of competitive/noncompetitive antagonists and enzyme inhibitors.

B. Rate Theory

The rate theory as proposed by Croxatto and Huidobro [191] and Paton et al. [192-194] states that pharmacological activity (response) is a function of and proportional to the rate of drug-receptor association and dissociation rather than the number of receptors occupied. As conceptually conceived, an agonist's (stimulant) rate of association-dissociation is extremely rapid, producing many impulses per unit time, whereas an antagonist, having strong receptor binding properties and a high association rate, is slow to dissociate. Thus the distinction between an agonist and an antagonist is solely dependent on k_2, high k_2 constants favoring agonist-type drugs while low k_2 values are normally associated with antagonists. Hence the dissociation constant became crucial: high dissociation rate implied high efficacy; low dissociation rate meant firm binding, prolonged occupancy, and antagonism; partial agonists had intermediate values of k_2 [167].

Since each association constitutes a "quantum" of stimulus, the association rate A is equivalent to

$$A = k_1[D](1 - p) \tag{189}$$

where

$[D]$ = drug concentration
p = fraction of receptor sites occupied
$1 - p$ = fraction of free receptors
k_1 = association constant

At equilibrium when the rates of association and dissociation are the same,

$$A = \frac{k_2}{1 + \{(k_1/k_2)/[D]\}} \tag{190}$$

As pointed out and stressed by Paton [167]:

> The most important theoretical point was that efficacy was quantized; i.e., an infinitely variable range of states of the receptor ranging from high to zero efficacy was replaced by two states (either free or bound, the binding being associated with release of a quantum of excitation), the infinite variation being in rate of change.

C. Induced-Fit Theory

The induced-fit theory of substrate-enzyme interaction as advanced by Koshland [195,196] in which a conformational change in the enzyme is induced by its combination with the substrate was proposed as an alternative model to account for a drug acting as agonist or antagonist at the molecular level. Concerned with the ability of proteins to distinguish between closely similar structures in such diverse biological phenomena as drug-receptor, hapten-antibody, substrate-enzyme, and inducer-induced enzyme systems, Koshland realized that the specificity patterns were sufficiently similar to each other to suggest that a mechanism with many common features operated in all of these systems. These distinguishing features that led to this conclusion were: (a) hydrophobic bonds, electrostatic bonds, hydrogen bonds, and chelation are the principal attractive forces; (b) electrostatic repulsions and steric hindrance are the principal repulsive forces; (c) a different combination of these forces is established in individual cases by means of the three-dimensional geometry of the protein; (d) the amino acid sequence of the protein plays a key role in establishing this three-dimensional geometry; (e) only a small fraction of the amino acid residues of the protein are directly involved in this specificity, but some that are distant from the "active site" play a role in establishing the necessary geometry of the essential amino acids; and (f) the manner in which the individual components combine to determine the specificity may include a dynamic interaction in which the small molecule induces a change in the protein conformation as a necessary part of the specificity process [195]. To explain the many inconsistencies of the "key-lock theory" for enzyme specificity (a theory that viewed the enzyme as a rigid negative of the substrate and stipulated that the substrate had to fit into this negative to react), Koshland suggested that these anomalies could be accounted for in the following manner: (a) a precise orientation of catalytic groups is required for enzyme action; (b) the substrate may cause an appreciable change in the three-dimensional relationship of the amino acids at the active site; and (c) the changes in protein structure caused by a substrate will bring the catalytic groups into the proper orientation for reaction, whereas a nonsubstrate will not. This set of postulates has been called the "induced-fit theory" for brevity and to emphasize that while the idea of a fit is retained from the key-lock theory, the fit in this case occurs only after the changes are induced by the substrate itself [186].

Originally proposed to explain enzymatic action on substrates, the induced-fit theory has been used to describe the consequences of drug-enzyme (receptor) interactions. However, as pointed out by Korolkovas [116];

> In conformity with the induced-fit theory, interaction of the drug should not necessarily be with an enzyme . . . it may be with a non-catalytic protein. If the interaction is with an enzyme, it may be not only at its active site but also at its allosteric site. By adapting the induced-fit theory to the rate theory, it could be stated that, if drug-receptor complexation results in such a conformation that a drug binds less strongly and dissociates more easily, it is an agonist; if, however, the complexation does not result in a change in conformation, the binding will be stable and an antagonist rather than an agonist is involved.

D. Macromolecular Perturbation Theory

The macromolecular perturbation theory was proposed by Belleau [197,198] in 1964 as a means for explaining drug action based on induced conformational perturbations of receptors. He proposed that interaction of substrate or drug of small molecular size with a macromolecule such as the protein of a drug receptor could lead to either a specific conformational perturbation (SCP) or to a nonspecific conformational perturbation (NSCP). Whereas a SCP change would result in a specific response derived from the drug-receptor complex possessing intrinsic activity, no pharmacological effect would be obtained from a nonspecific conformational perturbation which essentially produces an antagonistic or blocking action. On the other hand, a partial agonistic action could result from a drug having features which contribute to the formation of an equilibrium mixture of both SCP and NSCP complexes.

The MPT concept for drug action assumes that the protein is transformed from its resting state P to a conformally rearranged active state P* when a small molecule M_s (structurally related to the natural regulator or neurotransmitter) form an addition complex with the regulatory site [198].

$$P + M_s \xrightleftharpoons{k_1} P^*M_s \tag{191}$$

In this specifically arranged form, the receptor catalyzes the transformation of substrate molecule A to B which is essential for the initiation of a stimulus.

$$P^*M_s + A \rightleftharpoons P^*M_sA \rightarrow P^*M_s + B \tag{192}$$

On the other hand, an unfavorable conformation $P^{\#}$ may result if the regulatory site interacts with another molecule, M_i. In this situation, competitive antagonism would result since the receptor would remain catalytically ineffective as a catalyst for the conversion of A to B

$$P + M_i \xrightleftharpoons{k_2} P^{\#}M_i \tag{193}$$

$$P^{\#}M_i + A \rightleftharpoons P^{\#}M_iA \not\rightarrow P^{\#}M_i + B \tag{194}$$

Whereas the M_s-type molecules would have an ordering effect on the protein relative to the efficiency parameter, M_i-type molecules (inhibitory in nature) would produce disordering or disorientation. Hence the conformational transition from an ordered to a disordered arrangement could either be very abrupt as would be anticipated for compounds of very different structure or gradual as in a homologous series. It can also be rationalized on a probability basis that the structural transition (e.g., in a homologous series) from a M_s-type to a M_i-type molecule could be rather gradual, thus giving rise to the possibility that some drug molecules (M_{si}) could transform P into an equilibrium mixture of P^* and $P^\#$ complexes via either of the two comparable energy paths for the reaction. When this possibility exists, these agents or compounds will act as partial agonists (stimulants). This ambivalent phenomenon is rationalized by

$$P^{\#}M_{si} \underset{k_4}{\rightleftharpoons} P + M_{si} \overset{k_3}{\rightleftharpoons} P^{*}M_{si} \qquad (195)$$

When k_3 is equal to zero, Eq. (193) will be applicable. On the other hand, when $k_4 = 0$, the process can be described by Eq. (191).

As stressed by Belleau, the above

> will be possible if the free energy changes accompanying the formation of two kinds of complexes are of comparable magnitudes. In such a case, increasing the concentration of the drug molecules will never lead to the observation of maximal responses. In Ariens' terminology, such drugs will exhibit an intrinsic activity smaller than unity or in Stephenson's terminology, they will behave as partial agonists. It is also possible that partial agonists do not stabilize maximally the transition state for the conversion of A to B, and it is in this sense that the analogy between intrinsic activity and the rate-limiting step of an enzyme catalyzed reaction would be justified in principle. However, the equilibrium mixture hypothesis [Eq. (195)] is preferred because partial agonists are molecules (M_{si}) which frequently share the structural features with both the M_s- and M_i-types of molecules. The observation of "bell-shaped" dose-response curves (i.e., with maxima) with certain drug molecules is a phenomenon which has its equivalent in the field of enzymology.

The various dose-response curves on the basis of Eqs. (191), (195), and (196) for agonists [curve (a)], partial agonists [curve (b)], and the ternary complex formation [curve (c)], respectively, are shown in Fig. 22. With M_s molecules, maximal response is always achieved since curve (a) is a simple reflection of the law of mass action. For different M_s molecules, the observed parallel

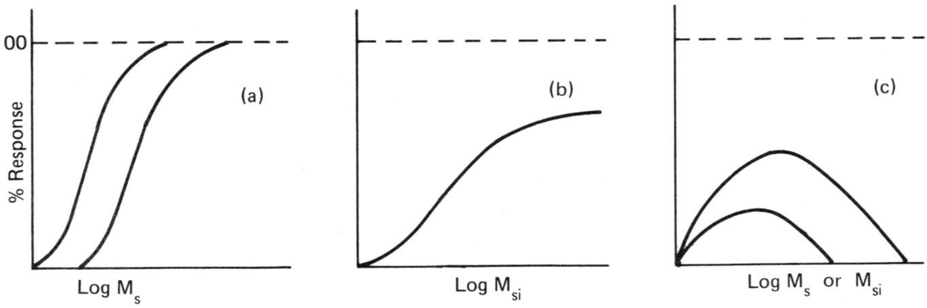

Figure 22 Dose-response curves derived from the macromolecular perturbation theory.

shift is an index of the relative efficiencies in the induction of a SCP in the protein or macromolecule. With M_i-type molecules, competitive antagonism will be noted as a consequence of the formation of a NSCP in the macromolecule [Eq. (193)]. However, with a M_{si}-type molecule, submaximal responses will be observed as a consequence of the formation of the equilibrium mixture of $P^{\#}M_{si}$ and $P^{*}M_{si}$ complexes [Eq. (195)]. Finally, a process such as that depicted by Eq. (196) at the receptor level,

$$P^{*}M_s \text{ or } P^{*}M_{si} + nM_s \text{ or } nM_{si} \rightleftharpoons P^{\#}(M_s)^n \text{ or } P^{\#}(M_{si})^n \qquad (196)$$

will be reflected by the bell-shaped dose-response curves such as those shown in Fig. 22c.

V. CHARACTERIZATION OF DRUG EFFECTS

A. Dose-Response Relationships

Ariens [199] views the dose-response relationship as the basis for routine therapy and the dose-effect curve as an instrument for establishing the relationship between the dose and the distribution of individuals in a population that reacts with a certain standard effect. However, the dose-effect curve is subject to many variables whose sequence of events can be considered separately under distinct phases:

1. The pharmaceutical phase, which includes the disintegration of the dosage form, the release of drug into solution from

the drug preparation, and possible drug loss due to inactivation before absorption—all important factors contributing to the pharmaceutical availability profile and the variability in the dose-effect relationship.

2. The pharmacokinetic phase of drug action, governed by such processes as absorption, distribution, drug binding, biotransformation, and excretion, which influence the drug levels achieved in the target tissue (hence its ultimate effect).
3. The pharmacodynamic phase, which is concerned with the drug-receptor interactions in the target compartment, where an induced stimulus results in the effect observed.

As a means for comparing and evaluating drug effectiveness and safety, it is customary to examine the magnitude of the drug response as a function of the administered dose. From the sigmoidal log dose-response curve, valuable drug information can be deduced: (a) maximum efficacy or intrinsic activity, (b) potency, (c) affinity, and (d) drug-receptor interaction attributed to either an agonist, partial agonist, or antagonist.

However, to describe the LDR curve on the basis of mass action as it applies to the occupancy theory of drug action, let us designate the fraction of the maximum response obtained at a given dose as f. According to Eq. (182), f can be expressed as

$$f = \frac{E}{E_{max}} = \frac{[DR]}{[R_{tot}]} \qquad (197)$$

and by substituting from Eq. (183) the term equivalent to E/E_{max} and then rearranging, Eq. (197) becomes

$$f = \frac{[D]}{[D] + K_D} \qquad (198)$$

or

$$[D] = K_D\left[\frac{f}{1 - f}\right] \qquad (199)$$

Since $[D_{tot}] - [DR] = [D]$ and $f[R_{tot}] = [DR]$ (Eq. (197)), it follows that the total drug may be expressed as

$$[D_{tot}] = K_D\left[\frac{f}{1 - f}\right] + f[R_{tot}] \qquad (200)$$

Total drug = free drug + bound drug

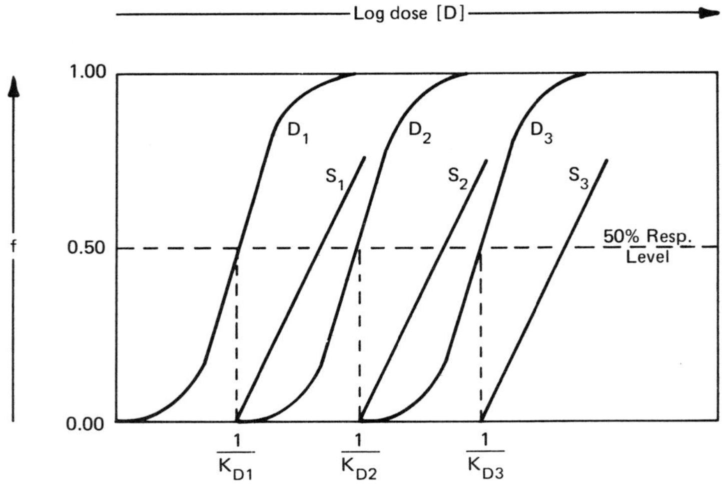

Figure 23 Log dose-response curves for drugs D_1, D_2, and D_3 (K_D and S notations explained in the text).

If, according to theory, (a) one receptor site combines with one drug molecule, (b) response is proportional to receptor occupancy, (c) all receptor sites are identical and readily accessible to the drug, and (d) only a negligible fraction of the total drug to which the receptors are exposed is combined so that the effective drug concentration is not altered during the reaction, the term $f[R_{tot}]$ can be ignored since it is considered to be much smaller than $K_D[f/(1 - f)]$. Hence

$$[D_{tot}] \cong [D] = K_D \frac{f}{1 - f} \tag{201}$$

As pointed out by Goldstein et al. [1], if $[R_{tot}]$ is much larger than K_D, the free drug term becomes negligible for any moderate value of f, and under these circumstances, where essentially all drug molecules are combined,

$$[D_{tot}] = f[R_{tot}] \tag{202}$$

Accordingly, if one plots f versus the logarithm of [D], a typical log dose-response curve such as that shown in Fig. 23 will be

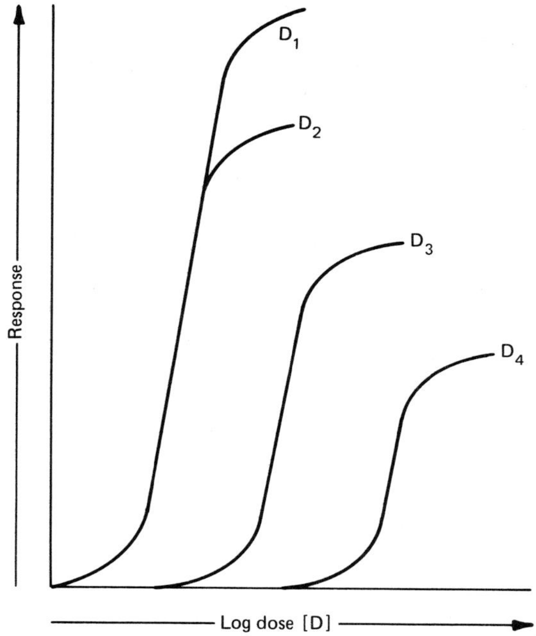

Figure 24 LDR curves for drugs with different affinities and efficacies: D_1 and D_2, same K_D but different efficacies; D_3 and D_4, different K_D and efficacy values.

obtained which is symmetrical about its midpoint, this midpoint representing the dose necessary to attain 50% maximal response when a perpendicular line is dropped from it to the abscissa (the horizontal axis with its log dose scale). For a congeneric series of drugs acting on a common receptor system, the maximal effect achieved for that system is assigned a value of 1.00 (or 100 if expressed as a percentage). Compounds that reach this maximal effect (1.00) are classified as full agonists (stimulants). The maximum effect produced by a drug is referred to as its efficacy. Thus efficacy (or intrinsic activity) describes the biologic effectiveness of the drug-receptor complex and was proposed as a modification to the occupancy theory of drug action to explain why different agonists apparently acting on the same receptor sites produced maximal responses of different magnitudes (Fig. 24). For the congeneric series, their LDR curves will appear as parallel sigmoidal curves with essentially parallel but invariant slopes at their midpoints. To produce effects of equal magnitude (i.e., at a 50% receptor occupancy level), the dissociation constant K_D is the determining factor as to the concentration of a particular drug that will be required to elicit this desired response. This K_D constant therefore may be considered as a measure of the drug's affinity for the receptor sites. The greater the value of K_D, the greater its affinity. Thus the position of each curve for a member of a congeneric series or compound acting on the same receptor site along the log dose axis is directly related to and determined by its K_D, this in turn being indicative of their relative affinities. The greater its K_D, the smaller will be the amount of drug required to produce the same intensity of response and, as evidenced by its relative position on the graph, it will be closer to the ordinate axis (Fig. 23).

The slope of the curve at the midpoint where it is most linear is found by differentiating the logarithmic form of Eq. (201) given by

$$\log [D] = \log K_D + \log\left[\frac{f}{1 - f}\right] \tag{203}$$

This differentiation proceeds as follows:

$$d \ln [D] = d \ln\left[\frac{1}{1 - f}\right] = \frac{df}{f(1 - f)} \tag{204}$$

$$\frac{df}{d \log [D]} = 2.303f(1 - f) \tag{205}$$

When f = 0.5 (at the midpoint),

$$\frac{df}{d \log [D]} = 0.576 \tag{206}$$

This may also be viewed as follows. When f = 0.5, the fraction f/(1 - f) will be equal to 1.0. Under these conditions, [D] = 1/K_D or K = 1/[D]. If a perpendicular is drawn from the point of 50% maximal response to the abscissa, its point of intersection gives the concentration of [D] equal to the reciprocal of the dissociation constant, 1/K_D, this again being an indication of its affinity for the receptor sites (Fig. 23).

As pointed out by Levine [4], potency, unlike affinity, is a comparative expression of drug activity, potency connoting the dose required to produce a particular effect relative to a given or implied standard of reference. It is essential to recognize that potencies are compared on the basis of doses that produce the same effect and not by comparing the magnitudes of effects elicited by the same dose [6]. Again, these relative potencies of drugs are reflected by the position of their LDR curves on the log dose axis. However, as stressed by Levine, two important points must be remembered: (a) the potency of a drug is in no way related to its value, efficacy, or safety; and (b) it is inappropriate to express the relative activities of two drugs in terms of potency unless they produce their effects by the same mechanism of action and can exert the same maximum effect.

Recall that agonist-receptor complex formation can be inhibited by other drugs known as competitive or noncompetitive antagonists. It is classified as competitive when it can be displaced by an excess of agonist from receptor sites for which it is "competing" with the agonist and with which it combines reversibly (i.e., the receptors). In this context, it can be said that the antagonist has affinity but lacks efficacy (intrinsic activity). From the agonist's LDR curve, one will note a parallel displacement to the right in the presence of a competitive antagonist (resulting from a change in the agonist's affinity for the receptor) without a change in either the curve's shape or maximum response (Fig. 25a).

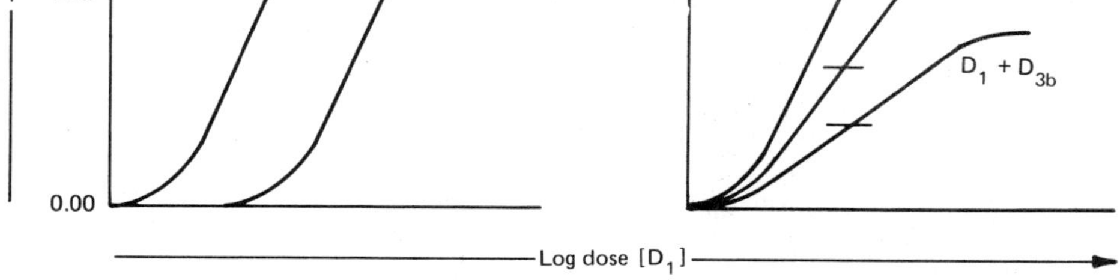

Figure 25 LDR curves depicting competitive (a) and noncompetitive (b) antagonism. (a) D_1, agonist; D_2, antagonist. (b) D_1, agonist; D_{3a} and D_{3b}, antagonist at two different concentration levels.

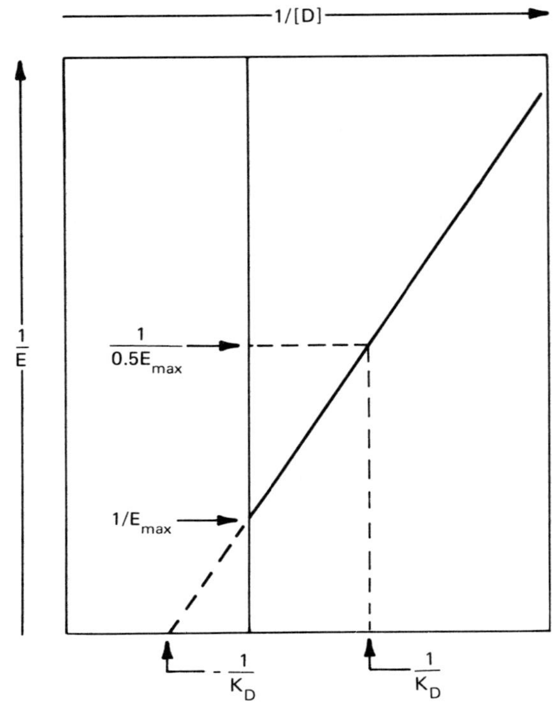

Figure 26 Double-reciprocal plot.

On the other hand, a noncompetitive antagonist produces different effects on the agonist's LDR curve that are readily recognized: (a) a decrease in its maximum height because of the inability of an excess of agonist to displace the antagonist from receptor sites, (b) a reduction of the slope of the curve in proportion to the magnitude of the noncompetitive antagonist produced, and (c) a concurrent shift to the right of the resulting agonist-antagonist curve (Fig. 25b). In addition to agonists and antagonists, some drugs are classified as partial antagonists that possess affinity and some efficacy and inhibit or prevent access of strong agonists to the receptor sites.

As noted previously, if one inverts Eq. (183), one obtains the equation for a straight line [Eq. (188)] when 1/E is plotted against 1/[D]. Known as a double-reciprocal plot, it can be used for the analysis of competitive/noncompetitive antagonists. Its slope will

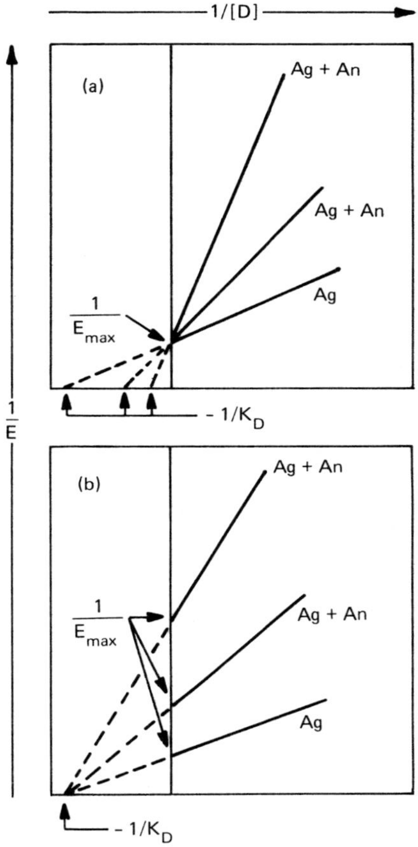

Figure 27 Use of double-reciprocal plots for the analysis of antagonisms: (a) competitive; (b) noncompetitive. Ag, agonist; An, antagonist.

be equivalent to K_D/E_{max} with the ordinate intercept given by $1/E_{max}$. On the other hand, K_D can be determined from the slope, by extrapolation to the abscissa from a point on the line equal to $1/0.5E_{max}$, or by extending the line to the abscissa, the intercept being equivalent to $-1/K_D$ as illustrated in Fig. 26. Using double-reciprocal plots, antagonisms can be analyzed as shown in Fig. 27. In Fig. 27b, a decrease in free receptors causes a proportional decrease in E_{max} when the agonist is subjected to the action of a noncompetitive antagonism. This will be accompanied by an increase in the ordinate intercept, but the abscissa intercept $-1/K_D$ will remain invariant. In the case of competitive antagonists, the apparent affinity of the agonist for the receptors will be reduced as reflected by the $-1/K_D$ intercept. Additionally, both the E_{max} and ordinate intercept $1/E_{max}$ will remain constant (Fig. 27a). The equations that can be used to describe the plots for noncompetitive and competitive antagonisms shown in Fig. 27 have been derived by Webb [115].

For a noncompetitive antagonist D_n which is capable of forming a complex D_nR (R being the receptor) with a dissociation constant K_{Dn}, its dissociation constant is

$$\frac{[D_n][R]}{[D_nR]} = K_{Dn} \tag{207}$$

Using the same approach as that for the derivation of Eq. (183), followed by substitution, one arrives at an expression such as

$$\frac{1}{E'} = \left(1 + \frac{[D_n]}{K_{Dn}}\right)\left(\frac{K_D}{E_{max}}\right)\frac{1}{[D]} + \left(1 + \frac{[D_n]}{K_{Dn}}\right)\frac{1}{E_{max}} \tag{208}$$

where E' is the magnitude of response in the presence of a fixed concentration of antagonist D_n. Compared to Eq. (188), both the slope and ordinate intercept ($1/E_{max}$) are increased by a factor equal to $[1 + (D_n/K_{Dn})]$. One can readily determine K_{Dn} since D_n is known.

For competitive antagonism, simultaneous equations must be solved since free receptor sites R are continuously at equilibrium with two complexes, DR and D_nR. Such being the situation, one obtains an equation such as the following:

$$\frac{1}{E'} = \left(1 + \frac{[D_n]}{K_{Dn}}\right)\left(\frac{K_{Dn}}{E_{max}}\right)\frac{1}{[D]} + \frac{1}{E_{max}} \tag{209}$$

As predicted, the ordinate intercept remains the same with only the slope increased by a factor of $[1 + (D_n/K_{Dn})]$. However, another change is noted. The original $-1/K_D$ intercept is also altered by the factor $[K_D/(K_{Dn} + D_n)]$.

In 1957, Gaddum [200] formulated the concept of dose ratio from Eqs. (198) and (210). The dose ratio $[D]_i/[D]$ is the ratio

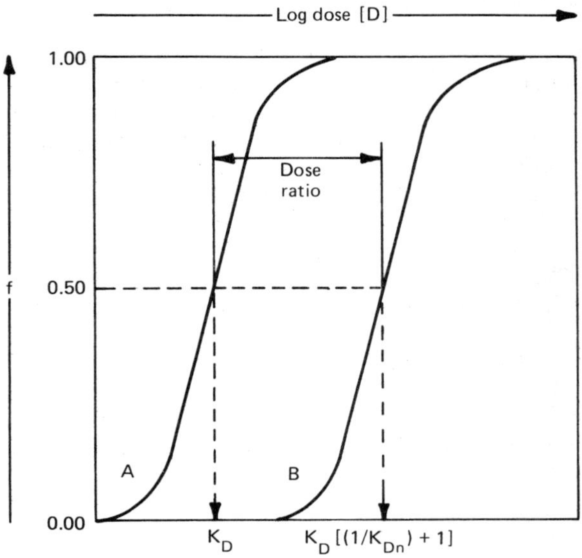

Figure 28 Graphical representation of the "dose ratio."

of two agonist concentrations producing the same response (i.e., the half-maximum response given by 0.5f) in the presence of $[D]_i$ and [D] of inhibitor (antagonist)

$$f = \frac{[D]}{[D] + K_D \left\{\left([D_n]/K_{Dn}\right) + 1\right\}} \tag{210}$$

According to Gaddum, the normal log dose-response curve is obtained as shown in Fig. 28 (curve A) when Eq. (198) is plotted. If Eq. (210) is plotted, it will yield a similar relationship in the presence of the antagonist (curve B). In contrast to curve A, whose position on the abscissa is fixed by the value of K_D, the position of curve B differs from that of A by a factor $([D_n]/K_{Dn}) + 1$. Since both curves are identically parallel, the distance between them $[([D_n]/K_{Dn}) + 1]$ is the ratio by which the original dose must be increased to obtain in the presence of antagonist a specified or desired control response. This can be expressed by

$$\frac{[D]_i}{[D]} - 1 = \frac{[D_n]}{K_{Dn}} \tag{211}$$

As noted by Paton [167]:

> The parallel shift and the $[([D]_i/[D]) - 1]$ relationship may be the two most important results of general pharmacological study of drug antagonism. The first is a test of competitive interaction which is independent of the shape of the dose-response curve or of the order of the reaction; thus it allowed experimental measures of competitive antagonism to be developed which can use the null method, estimating dose ratios without having to make assumptions about the slope of the dose-response curves or being constrained as to the actual level of response used. The $[([D]_i/[D]) - 1]$ relationship provides a test for reaction order, for agonist as well as antagonist, as well as a further test of the whole approach.

For higher orders of agonist-antagonist interaction, the dose ratio may be expressed as shown below. Let us suppose that the following reactions occur between the agonist, antagonist, and receptor:

$$aD + R \rightleftharpoons aDR \qquad (212)$$

$$bD_n + R \rightleftharpoons bD_nR \qquad (213)$$

Under these circumstances, the dose ratio is given by

$$\left(\frac{[D]_i}{[D]}\right)^a = \frac{[D_n]^b}{K_{Dn}} + 1 \qquad (214)$$

B. Response-Time or Dose-Time Relationships

As indicated in Fig. 1, there are many factors that influence the onset, duration, and intensity of drug effects. The amount of drug actually reaching the site of action (receptor site) is governed by many simultaneously operating processes: (a) the extent and rate of drug absorption from its site of administration, which, in turn, is influenced by the manner or route in which the drug is introduced into the body; (b) the rate at which the drug is distributed throughout the body, particularly to the site of action; (c) the rate and extent of drug biotransformation; (d) the processes associated with drug-binding and storage in various body depots; and (e) the rate of excretion. Many of these

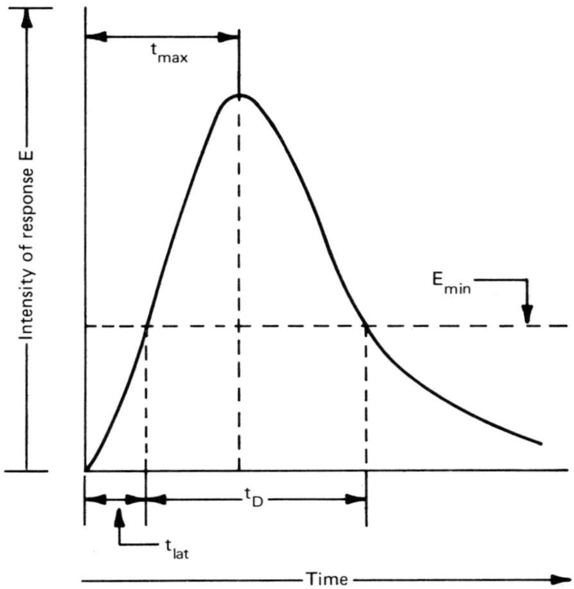

Figure 29 Drug response as a function of time.

processes, in addition to drug-response considerations, have been treated mathematically in this text. Since these are all rate functions, it is possible to obtain some valuable information about these complex processes (not individually, but combined) which regulate the observed dose-response relationships by monitoring the magnitude of the drug-response or effect as a function of time after single dose administration.

The time course of drug effect is commonly characterized by three parameters: (a) latency, (b) time of peak effect, and (c) duration of action. Latency is the time t_{lat} between drug administration and the onset or minimum level of measurable response E_{min} as shown in Fig. 29. The time of peak effect t_{max} defines that time at which the drug level is maximum. The duration of drug action t_D is that time period extending from the first measurable response (onset) until drug elimination is such that its concentration falls below that which is required to produce an observed response (E_{min}).

If it is assumed that a correlation exists between a drug's blood concentration and its pharmacological response, its duration of action can be established by determining its blood levels at various times after administration. In this manner, a profile of the drug level versus time can be derived (hence pharmacological response versus time).

If the absorption and elimination rate processes are first-order, the time of peak effect t_{max} can be determined by Eq. (45) for the situation where $k_a \neq k_e$ (i.e., provided that both rate constants are known). When $k_a = k_e$, Eq. (46) is used. Regardless of the dose administered, Eqs. (45) and (46) indicate that the time in which the maximum drug level is achieved is constant. On the other hand, the maximum amount of drug in the body can be estimated by substituting values of t_{max} into the original equations. For the situation where $k_a \neq k_e$ or $k_a = k_e$, these maximum amounts are given by Eqs. (47) and (48), respectively. Additionally, drug plasma concentration C_p for a single-compartment open model to describe the kinetics of drug levels after single doses can be estimated by Eq. (107), whereas its concentration in the central compartment of a two-compartment model at time t after a single dose is given by Eq. (108).

To establish the minimum level of drug required in the body at the site of action which will yield a measurable minimum response (E_{min}), this minimum effective dose value can be obtained from a semilogarithmic plot of dose administered versus duration of effect (response), as shown in Fig. 30. This minimum effective dose level is that amount of drug necessary to produce the minimum measurable response (E_{min}). In Fig. 30, this minimum dose is identified as D_{min}.

If it is assumed that the rate constant k_r for the removal of a drug from a receptor is a first-order constant, the duration of effect or response t_D is given by

$$D_{min} = D_0 e^{-k_r t_D} \qquad (215)$$

or

$$t_D = \frac{2.30}{k_r} \log \frac{D_0}{D_{min}} \qquad (216)$$

Thus one notes that the therapeutic level of a drug depends on the ratio of the administered dose D_0 to the minimum effective dose D_{min}. In Eq. (216), k_r may be estimated from a plot such as that shown in Fig. 30 since values for the slope $2.30/k_r$ as well as D_{min} can be obtained. As pointed out by Goldstein et al. [1], if

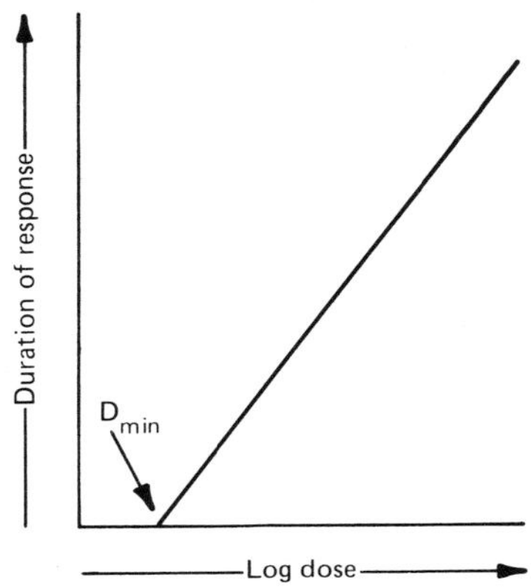

Figure 30 Relationship between dose administered and duration of response.

twice the minimum effective dose is administered, the duration of therapeutic level will be simply the biologic half-life.

Knowledge of minimum effective dose, minimum measurable response, duration of drug effect, time of peak effect, and biologic half-life is extremely important for establishing dose regimens. The details of the mathematics involved for calculating the time at which maximum accrual of drug will occur on a fixed-dose schedule are adequately covered by Goldstein et al. [1] and Rowland [45] in their discussions of drug administration and regimens.

References

1. Goldstein, A., Aronow, L., and Kalman, S. M., *Principles of Drug Action: The Basis of Pharmacology*, 2nd ed., Wiley, New York, 1974.
2. Goodman, L. S., and Gilman, A. (Eds.), *The Pharmacological Basis of Therapeutics*, 4th ed., Macmillan, New York, 1970.
3. Wilson, C. O., Gisvold, O., and Doerge, R. F. (Eds.), *Textbook of Organic Medicinal and Pharmaceutical Chemistry*, 6th ed., Lippincott, Philadelphia, 1971.
4. Levine, R. R., *Pharmacology: Drug Actions and Reactions*, Little, Brown, Boston, 1973.
5. LaDu, B. N., Mandel, H. G., and Way, E. L. (Eds.), *Fundamentals of Drug Metabolism and Drug Disposition*, Williams & Wilken, Baltimore, 1971.
6. Goth, A., *Medical Pharmacology: Principles and Concepts*, 7th ed., Mosby, St. Louis, 1974.
7. Ariens, E. J. (Ed.), *Molecular Pharmacology: The Mode of Action of Biologically Active Compounds*, Vol. 1, Academic Press, New York, 1964.
8. Crossland, J. (Ed.), *Lewis's Pharmacology*, 4th ed., Livingstone, Edinburgh, 1970.
9. Notari, R. E., *Biopharmaceutics and Pharmacokinetics*, Marcel Dekker, New York, 1971.

10. Tedeschi, D. H., and Tedeschi, R. E. (Eds.), *Importance of Fundamental Principles in Drug Evaluation*, Raven Press, New York, 1968.
11. Melmon, K. L., and Morrelli, H. F. (Eds.), *Clinical Pharmacology: Basic Principles in Therapeutics*, Macmillan, New York, 1972.
12. Csaky, T. Z., *Introduction to General Pharmacology*, Appleton-Century-Crofts, New York, 1969.
13. Kalow, W., *Pharmogenetics, Heredity and the Response to Drugs*, Saunders, Philadelphia, 1962.
14. Fulton, W. F. M. (Ed.), *Modern Trends in Pharmacology and Therapeutics*, Butterworth, London, 1967.
15. Schueler, F. W., *Chemobiodynamics and Drug Design*, McGraw-Hill, New York, 1960.
16. Papper, E. M., and Kitz, R. J. (Eds.), *Uptake and Distribution of Anesthetic Agents*, McGraw-Hill, New York, 1963.
17. Stein, W. D., *The Movement of Molecules Across Cell Membranes*, Academic Press, New York, 1967.
18. Fishman, W. H., *Chemistry of Drug Metabolism*, Thomas, Springfield, Ill., 1961.
19. Parke, D. V., *The Biochemistry of Foreign Compounds*, Pergamon Press, Elmsford, N.Y., 1968.
20. Meyers, F. H., Jawetz, E., and Goldfien, A., *Review of Medical Pharmacology*, 4th ed., Lange, Los Altos, Calif., 1974.
21. Binns, T. B. (Ed.), *Absorption and Distribution of Drugs*, Livingstone, Edinburgh, 1964.
22. Williams, R. T., *Detoxication of Drugs, Toxic Substances and Other Organic Compounds*, 2nd ed., Wiley, New York, 1959.
23. Laurence, D. R. (Ed.), *Quantitative Methods in Human Pharmacology and Therapeutics*, Pergamon Press, Oxford, 1959.
24. Goldstein, A., *Biostatistics: An Introductory Text*, Macmillan, New York, 1964.
25. Hill, A. B., *Principles of Medical Statistics*, 9th ed., Oxford University Press, London, 1971.
26. Finney, D. J., *Statistical Methods in Biological Assay*, 2nd ed., Griffin, London, 1964.
27. Daniel, W. W., *Biostatistics: A Foundation for Analysis in the Health Sciences*, Wiley, New York, 1974.
28. DiPalma, J. R. (Ed.), *Pharmacology in Medicine*, 4th ed., McGraw-Hill, New York, 1971.

29. Garattini, S., and Shore, P. A. (Eds.), *Advances in Pharmacology and Chemotherapy*, Vols. 1-11, Academic Press, New York, 1962-1973.
30. Ellis, G. P., and West, W. B. (Eds.), *Progress in Medicinal Chemistry*, Vols. 1-9, Butterworth, London, 1961-1973.
31. Sollman, T., *A Manual of Pharmacology*, 8th ed., Saunders, Philadelphia, 1957.
32. Gudzinowicz, B. J., Gudzinowicz, M. J., and Martin, H. F., *Fundamentals of Integrated GC-MS, Part I: Gas Chromatography: Basic Theory and Principles*, Marcel Dekker, New York, 1976.
33. Gudzinowicz, B. J., Gudzinowicz, M. J., and Martin, H. F., *Fundamentals of Integrated GC-MC, Part II: Mass Spectrometry*, Marcel Dekker, New York, 1976.
34. Gudzinowicz, B. J., Gudzinowicz, M. J., and Martin, H. F., *Fundamentals of Integrated GC-MS, Part III: The Integrated GC-MS Analytical System*, Marcel Dekker, New York, 1977.
35. Gudzinowicz, B. J., and Gudzinowicz, M. J., *Analysis of Drugs and Metabolites by Gas Chromatography-Mass Spectrometry*, Vol. 1: *Respiratory Gases, Volatile Anesthetics, Ethyl Alcohol, and Related Toxicological Materials*, Marcel Dekker, New York, 1977.
36. Gudzinowicz, B. J., and Gudzinowicz, M. J., *Analysis of Drugs and Metabolites by Gas Chromatography-Mass Spectrometry*, Vol. 2: *Hypnotics, Anticonvulsants, and Sedatives*, Marcel Dekker, New York, 1977.
37. Gudzinowicz, B. J., and Gudzinowicz, M. J., *Analysis of Drugs and Metabolites by Gas Chromatography-Mass Spectrometry*, Vol. 3: *Antipsychotic, Antiemetic, and Antidepressant Drugs*, Marcel Dekker, New York, 1977.
38. Gudzinowicz, B. J., and Gudzinowicz, M. J., *Analysis of Drugs and Metabolites by Gas Chromatography-Mass Spectrometry*, Vol. 4: *Central Nervous System Stimulants*, Marcel Dekker, New York, 1978.
39. Gudzinowicz, B. J., and Gudzinowicz, M. J., *Analysis of Drugs and Metabolites by Gas Chromatography-Mass Spectrometry*, Vol. 5: *Analgesics, Local Anesthetics, and Antibiotics*, Marcel Dekker, New York, 1978.
40. Gudzinowicz, B. J., and Gudzinowicz, M. J., *Analysis of Drugs and Metabolites by Gas Chromatography-Mass Spectrometry*, Vol. 6: *Cardiovascular, Antihypertensive, Hypoglycemic, and Thyroid-Related Agents*, Marcel Dekker, New York, 1979.

41. Gudzinowicz, B. J., and Gudzinowicz, M. J., *Analysis of Drugs and Metabolites by Gas Chromatography-Mass Spectrometry*, Vol. 7: *Natural, Pyrolytic, and Metabolic Products of Tobacco and Marijuana*, Marcel Dekker, New York, 1980.
42. Speizer, F. E., Doll, R., Heaf, P., and Strong, L. B., *Br. Med. J. 1*: 339 (1968).
43. Schanker, L. S., in *Fundamentals of Drug Metabolism and Drug Disposition*, LaDu, B. N., Mandel, H. G., and Way, E. L. (Eds.), Williams & Wilkins, Baltimore, 1971.
44. Levy, G., in *Physico-Chemical Aspects of Drug Actions (Proceedings of the Third International Pharmacological Meeting*, 1966, Vol. 7), Ariens, E. J. (Ed.), Pergamon Press, Oxford, 1968.
45. Rowland, M., in *Clinical Pharmacology: Basic Principles in Therapeutics*, Melmon, K. L., and Morrelli, H. F., (Eds.), Macmillan, New York, 1972.
46. Barr, W. H., *Am. J. Pharm. Educ. 52*: 958 (1968).
47. Robinson, D. S., *Postgrad. Med. 57*: 55 (1975).
48. Albert, A., and Serjeant, E. D., *Ionization Constants of Acids and Bases: A Laboratory Manual*, Wiley, New York, 1962.
49. Parke, T. V., and Davis, W. W., *Anal. Chem. 26*: 642 (1954).
50. Hogben, C. A. M., Tocco, D. J., Brodie, B. B., and Schanker, L. S., *J. Pharmacol. Exp. Ther. 125*: 275 (1959).
51. Schedl, H. P., and Clifton, J. A., *Gastroenterology 41*: 491 (1961).
52. Hansch, C., and Fujita, T., *J. Am. Chem. Soc. 86*: 1616 (1964).
53. Hammett, L. P., *Physical Organic Chemistry*, McGraw-Hill, New York, 1940.
54. Jaffe, H. H., *Chem. Rev. 53*: 191 (1953).
55. Taft, R. W., *J. Am. Chem. Soc. 75*: 4231 (1953).
56. Taft, R. W., in *Steric Effects in Organic Chemistry*, Newman, M. S. (Ed.), Wiley, New York, 1956.
57. Bellamy, L. J., *The Infrared Spectra of Complex Molecules*, 2nd ed., Methuen, London, 1958.
58. Butler, T. C., *J. Pharmacol. Exp. Ther. 108*: 474 (1953).
59. Ruhland, W., and Hoffmann, C., *Planta 1:* 1 (1925).
60. Collander, R., and Barlund, H., *Acta Bot. Fenn. 11*: 1 (1933).

61. Cohn, V. H., in *Fundamentals of Drug Metabolism and Drug Disposition*, LaDu, B. N., Mandel, H. G., and Way, E. L. (Eds.), Williams & Wilkins, Baltimore, 1971.
62. Teorell, T., *Arch. Int. Pharmacodyn.* *57*: 205, 226 (1937).
63. Fingl, E., and Woodbury, D. M., in *The Pharmacological Basis of Therapeutics*, 4th ed., Goodman, L. S., and Gilman, A. (Eds.), Macmillan, New York, 1970.
64. Butler, T. C., in *Fundamentals of Drug Metabolism and Drug Disposition*, LaDu, B. N., Mandel, H. G., and Way, E. L. (Eds.), Williams & Wilkins, Baltimore, 1971.
65. Chiou, W. L., *Res. Commun. Clin. Pathol. Pharmacol.* *7*: 679 (1974).
66. Perrier, D., and Gibaldi, M., *J. Pharmacokinet. Biopharm.* *1*: 17 (1973).
67. Rowland, M., in *Current Concepts in the Pharmaceutical Sciences*, Swarbrick, J. (Ed.), Lea & Febiger, Philadelphia, 1973.
68. Guyton, A. C., *Textbook of Medical Physiology*, 4th ed., Saunders, Philadelphia, 1971.
69. Spector, W. S., *Handbook of Biological Data*, Saunders, Philadelphia, 1956.
70. Gourley, D. R. H., in *Modern Trends in Pharmacology and Therapeutics*, Fulton, W. F. M. (Ed.), Butterworth, London, 1967.
71. Daniels, T. C., and Jorgensen, E. C., in *Textbook of Organic Medicinal and Pharmacological Chemistry*, 6th ed., Wilson, C. O., Gisvold, O., and Doerge, R. F. (Eds.), Lippincott, Philadelphia, 1971.
72. Brodie, B. B., Mark, L. C., Papper, E. M., Lief, E. B., Bernstein, E., and Rovenstine, E. A., *J. Pharmacol. Exp. Ther.* *98*: 85 (1950).
73. Brodie, B. B., Bernstein, E., and Mark, L. C., *J. Pharmacol. Exp. Ther.* *105*: 421 (1952).
74. Axelrod, J., Aranow, L., and Brodie, B. B., *J. Pharmacol. Exp. Ther.* *106*: 166 (1952).
75. Brodie, B. B. Aronow, L., and Axelrod, J., *J. Pharmacol. Exp. Ther.* *111*: 21 (1954).
76. Agarwal, S. L., and Harvey, S. C., *J. Pharmacol. Exp. Ther.* *117*: 106 (1956).
77. Brodie, B. B., and Hogben, C. A. M., *J. Pharm. Pharmacol.* *9*: 345 (1957).
78. Welch, K., and Friedman, V., *Brain* *83*: 454 (1960).

79. Davison, C., in *Fundamentals of Drug Metabolism and Drug Disposition*, LaDu, B. N., Mandel, H. G., and Way, E. L. (Eds.), Williams & Wilkins, Baltimore, 1971.
80. Hansten, P. D., *Drug Interactions*, 2nd ed., Lea & Febiger, Philadelphia, 1973.
81. Nozaki, Y., Gurd, F. R. N., Chen, C. F., and Edsall, J. T., *J. Am. Chem. Soc.* *79*: 2123 (1957).
82. Scatchard, G., Scheinberg, I. H., and Armstrong, S. H., *J. Am. Chem. Soc.* *72*: 535 (1950).
83. Klotz, I. M., Walker, F. M., and Pivan, R. B., *J. Am. Chem. Soc.* *68*: 1486 (1946).
84. Rehberg, P. B., *Acta Physiol. Scand.* *5*: 305 (1943).
85. Potter, G. D., and Guy, J. L., *Proc. Soc. Exp. Biol. Med.* *116*: 658 (1964).
86. Hummel, J. P., and Dreyer, W. J., *Biochem. Biophys. Acta* *63*: 530 (1962).
87. Chignell, C. F., *Adv. Drug Res.* *5*: 55 (1970).
88. Klotz, I. M., Burkhard, R. K., and Urquhart, J. M., *J. Am. Chem. Soc.* *74*: 202 (1952).
89. Tata, J. R., Windnell, C. C., and Gratzer, W. B., *Clin. Chem. Acta* *6*: 597 (1961).
90. Bickel, M. H., and Bovet, D., *J. Chromatogr.* *8*: 466 (1962).
91. Scatchard, G., *Ann. N.Y. Acad. Sci.* *51*: 660 (1949).
92. Scholtan, W., *Arzneim.-Forsch.* *14*: 1139 (1964).
93. Muller, W. E., and Wollert, U., *Res. Commun. Chem. Pathol. Pharmacol.* *9*: 413 (1974).
94. Chiou, W. L., and Hsiao, J. H., *Res. Commun. Chem. Pathol. Pharmacol.* *8*: 273 (1974).
95. Newton, B. A., *Adv. Pharmacol.* *8*: 150 (1970).
96. Reich, E., *Science* *143*: 684 (1964).
97. Hamilton, L. D., Fuller, W., and Reich, E., *Nature* *198*: 538 (1963).
98. Muller, W., and Crothers, D. M., *J. Mol. Biol.* *35*: 251 (1968).
99. Hahn, F. E., O'Brien, R. L., Ciak, J., Allison, J. L., and Olenick, J. G., *Mil. Med.* *131*: 1071 (1966).
100. Parker, F. S., and Irvin, J. L., *J. Biol. Chem.* *199*: 897 (1952).
101. Kurnick, N. B., and Radcliffe, I. E., *J. Lab. Clin. Med.* *60*: 669 (1962).
102. Potts, A. M., *Invest. Ophthalmol.* *3*: 405 (1964).
103. Potts, A. M., *Invest. Ophthalmol.* *1*: 522 (1962).

104. Bernstein, H., Zvaifler, N., Rubin, M., and Mansour, A. M., *Invest. Ophthalmol. 2*: 384 (1963).
105. Goldstein, A., *Pharm. Rev. 1*: 102 (1949).
106. Borga, O., Azarnoff, D. L., and Sjoeqvist, F., *J. Pharm. Pharmacol. 20*: 571 (1968).
107. Sturman, J. A., and Smith, M. G. H., *J. Pharm. Pharmacol. 19*: 621 (1967).
108. Starr, P., Nicoloff, J. T., and Pilleggi, V. J., *Clin. Res. 15*: 127 (1967).
109. Mandel, H. G., in *Fundamentals of Drug Metabolism and Drug Disposition*, LaDu, B. N., Mandel, H. G., and Way, E. L. (Eds.), Williams & Wilkins, Baltimore, 1971.
110. Brodie, B. B., Gillette, J. R., and LaDu, B. N., *Annu. Rev. Biochem. 27*: 427 (1958).
111. Kalow, W., *Proc. Second Int. Pharm. Mtg. 4*: 245 (1965).
112. Gillette, J. R., *Prog. Drug Res., 6*, 13 (1963).
113. Malmstrom, B. G., and Rosenberg, A., *Adv. Enzymol. 21*: 131 (1959).
114. Dixon, M., and Webb, E. C., *Enzymes*, 2nd ed., Academic Press, New York, 1964.
115. Webb, J. L., *Enzyme and Metabolic Inhibitors*, Academic Press, New York, 1963.
116. Korolkovas, A., *Essentials of Molecular Pharmacology: Background for Drug Design*, Wiley-Interscience, New York, 1970.
117. Gillette, J. R., in *Fundamentals of Drug Metabolism and Drug Disposition*, LaDu, B. N., Mandel, H. G., and Way, E. L. (Eds.), Williams & Wilkins, Baltimore, 1971.
118. Westley, J., *Enzymatic Catalysis*, Harper & Row, New York, 1969.
119. LaDu, B. N., *Hospital Practice*, June 1971.
120. Gillette, J. R., *Prog. Drug Res., 6*, 49 (1963).
121. Shideman, F. E., and Mannering, G. J., *Annu. Rev. Pharmacol. 3*: 33 (1963).
122. Brodie, B. B., *Proc. Mid-Year Am. Pharm. Mfg. Assoc. Mtg.*, 1952.
123. Smith, R. L., and Williams, R. T., in *Glucuronic Acid*, Dutton, G. J. (Ed.), Academic Press, New York, 1966.
124. Boyland, E., and Ramsay, G. S., *Biochem. J. 77*: 182 (1960).
125. Wollman, H., and Dripps, R. D., in *The Pharmacological Basis of Therapeutics*, 4th ed., Goodman, L. S., and Gilman, A. (Eds.), Macmillan, New York, 1970.

126. Butler, T. C., *Fed. Proc. 17*: 1158 (1958).
127. Butler, T. C., *Proc. First Int. Pharm. Mtg. 6*: 193 (1962).
128. Cafruny, E. J., in *Fundamentals of Drug Metabolism and Drug Disposition*, LaDu, B. N., Mandel, H. G., and Way, E. L. (Eds.), Williams & Wilkins, Baltimore, 1971.
129. Williams, R. T., Millburn, P., and Smith, R. L., *Ann. N.Y. Acad. Sci. 123*: 110 (1965).
130. Cox, E., and Wright, S. E., *J. Pharmacol. Exp. Ther. 126*: 117 (1959).
131. Haddock, E. P., Zapolski, E. J., Rubin, M., and Princiotto, J. V., *Proc. Soc. Exp. Biol. Med. 120*: 663 (1965).
132. Meyer-Brunot, H. G., and Keberle, H., *Am. J. Physiol. 214*: 1193 (1968).
133. Stowe, C. M., and Plaa, G. L., *Annu. Rev. Pharmacol. 8*: 337 (1968).
134. Golman, K., *J. Pharm. Sci. 64*: 405 (1975).
135. Nayak, R. K., Smyth, R. D., Chamberlain, J. H., Polk, A., De Long, A. F., Herczeg, T., Chemburkar, P. B., Joslin, R. S., and Reavey-Cantwell, N. H., *J. Pharmacokinet. Biopharm. 2*: 107 (1974).
136. Wagner, J. G., *Biopharmaceutics and Relevant Pharmacokinetics*, Hamilton Press, Hamilton, Ohio, 1971.
137. Kramer, W. G., Lewis, R. P., Cobb, T. C., Forester, W. F., Jr., Visconti, J. A., Wanke, L. A., Boxenbaum, H. G., and Reuning, R. H., *J. Pharmacokinet. Biopharm. 2*: 299 (1974).
138. Benet, L. Z., *J. Pharm. Sci. 61*: 536 (1972).
139. Weast, R. C. (Ed.), *Handbook of Tables for Mathematics*, 4th ed., Chemical Rubber Co., Cleveland, 1970.
140. Mihailova, D., Rosen, A., Testa, B., and Beckett, A. J., *J. Pharm. Pharmacol. 26*: 711 (1974).
141. Chiou, W. L., and Hsu, F. H., *Res. Commun. Chem. Pathol. Pharmacol. 10*: 315 (1975).
142. Tozer, T. N., *J. Pharmacokinet. Biopharm. 2*: 13 (1974).
143. Kaplan, S. A., in *Current Concepts in the Pharmaceutical Sciences*, Swarbrick, J. (Ed.), Lea & Febiger, Philadelphia, 1973.
144. Chiou, W. L., *J. Clin. Pharmacol. 14*: 418 (1974).
145. Perrier, D., and Gibaldi, M., *J. Clin. Pharmacol. 20*: 144 (1973).
146. Enger, E., and Blegen, E. M., *Scand. J. Clin. Lab. Invest. 16*: 273 (1964).
147. Graham, G., and Bolt, A. G., *J. Pharmacokinet. Biopharm. 2*: 191 (1974).

148. Guggenheim, E. A., *Phil. Mag. 2*: 538 (1926).
149. Beckett, A. H., and Rowland, M., *J. Pharm. Pharmacol. 17*: 628 (1965).
150. Anfinsen, C. B., *The Molecular Basis of Evolution*, Wiley, New York, 1959.
151. Albert, A., *Selective Toxicity*, 3rd ed., Wiley, New York, 1965.
152. Karnofsky, D. A., *Ann. N.Y. Acad. Sci. 68*: 657 (1958).
153. Brown, S. S., *Adv. Pharmacol. 2*: 243 (1963).
154. Brookes, P., and Lawley, P. D., *Br. Med. Bull. 20*: 91 (1964).
155. Davies, D. R., and Green, A. L., *Adv. Enzymol. 20*: 283 (1958).
156. Wilson, I. B., *Fed. Proc. 18*: 752 (1959).
157. Casida, J. E., *Science 146*: 1011 (1964).
158. Sprague, J. M., *Top. Med. Chem. 2*: 1 (1968).
159. Watson, J. H., *Molecular Biology of the Gene*, W. A. Benjamin, New York, 1965.
160. Baker, B. R., *Design of Active-Site-Directed Irreversible Enzyme Inhibitors*, Wiley, New York, 1967.
161. Reutov, O., *The Theoretical Principles of Organic Chemistry*, Mir, Moscow, 1967.
162. Gill, E. W., *Prog. Med. Chem. 4*: 39 (1965).
163. Barrow, G. M., *Physical Chemistry*, 2nd ed., McGraw-Hill, New York, 1966.
164. Bernhard, S. A., *The Structure and Function of Enzymes*, W. A. Benjamin, New York, 1968.
165. Mantione, M. J., in *Molecular Association in Biology*, Pullman, B. (Ed.), Academic Press, New York, 1968.
166. Stephenson, R. P., *Br. J. Pharmacol. 11*: 379 (1956).
167. Paton, W. D. M., in *Molecular Properties of Drug Receptors*, Porter, P., and O'Connor, M. (Eds.), Churchill, London, 1970.
168. Sexton, W. A., *Chemical Constitution and Biological Activity*, 3rd ed., Spon, London, 1963.
169. Beckett, A. H., *Prog. Drug Res. 1*: 455 (1959).
170. Ferguson, J., *Proc. R. Soc. (Lond.), Ser. B, 127*: 387 (1939).
171. Burger, A., in *Medicinal Chemistry*, 3rd ed., Burger, A. (Ed.), Wiley-Interscience, New York, 1970.
172. Robinson, N., *J. Pharm. Pharmacol. 12*: 129, 193 (1960).
173. Fastier, F. N., *Annu. Rev. Pharmacol. 4*: 51 (1964).
174. Burger, A., and Parulkar, A. P., *Annu. Rev. Pharmacol. 6*: 19 (1966).

175. Van Possum, J. B., *J. Pharm. Pharmacol. 15*: 285 (1963).
176. Albert, A., *Annu. Rev. Pharmacol. 11*: 13 (1971).
177. Cavallite, C. J., *Annu. Rev. Pharmacol. 8*: 39 (1968).
178. Ariens, E. J., *Prog. Drug Res. 10*: 429 (1966).
179. Ariens, E. J., *Farm. Ed. Sci. (Pavia), 23*: 52 (1968).
180. Langmuir, I., *J. Am. Chem. Soc. 41*: 868, 1543 (1919).
181. Friedman, H. L., in *First Symposium on Chemical-Biological Correlation*, May 1950, NAS-NRC Publ. 206, Washington, D.C., 1951.
182. Harper, N. J., *Prog. Drug Res. 4*: 221 (1962).
183. Easson, L. H., and Stedman, E., *Biochem. J. 27*: 1257 (1933).
184. Clark, A. J., *J. Physiol. 61*: 530, 547 (1926).
185. Gaddum, J. H., *J. Physiol. 61*: 141 (1926).
186. Gaddum, J. H., *J. Physiol. 89*: 7P (1937).
187. Ariens, E. J., *Arch. Int. Pharmacodyn. 99*: 32 (1954).
188. Ariens, E. J., and Simonis, A. M., *J. Pharm. Pharmacol. 16*: 137, 289 (1964).
189. Ariens, E. J., van Rossum, J. M., and Koopman, P. C., *Arch. Int. Pharmacodyn. 127*: 459 (1960).
190. Ariens, E. J., van Rossum, J. M., and Koopman, P. C., *Arch. Int. Pharmacodyn. 136*: 385 (1962).
191. Croxatto, R., and Huidobro, F., *Arch. Int. Pharmacodyn. 106*: 207 (1956).
192. Paton, W. D. M., *Proc. R. Soc. (Lond.), Ser. B, 154*: 21 (1961).
193. Paton, W. D. M., and Rang, H. P., *Adv. Drug Res. 3*: 57 (1966).
194. Paton, W. D. M., and Payne, J. P., *Pharmacological Principles and Practice*, Churchill, London, 1968.
195. Koshland, D. E., Jr., *Biochem. Pharmacol. 8*: 57 (1961).
196. Koshland, D. E., Jr., *Proc. Natl. Acad. Sci. USA 44*: 98 (1958).
197. Belleau, B., *J. Med. Chem. 7*: 776 (1964).
198. Belleau, B., in *Advances in Drug Research*, Vol. 2, Harper, N. J., and Simmonds, A. B. (Eds.), Academic Press, London, 1965.
199. Ariens, E. J., *Clin. Pharmacol. Ther. 16*: 155 (1974).
200. Gaddum, J. H., *Pharmacol. Rev. 9*: 211 (1957).

Author Index

Numbers in parentheses are reference numbers and indicate that an author's work is referred to. Numbers in italics give the page on which the complete reference is listed.

Subject Index

D

E

M

Q